Adventures *in* Volcanoland

Tamsin Mather

Adventures *in* Volcanoland

What Volcanoes Tell Us
About the World and Ourselves

abacus
books

ABACUS

First published in Great Britain in 2024 by Abacus

1 3 5 7 9 10 8 6 4 2

Maps by Stephen Dew

A CIP catalogue record for this book
is available from the British Library.

Hardback ISBN 978-1-4087-1461-4
C-format ISBN 978-1-4087-1462-1

Typeset in Jenson by M Rules
Printed and bound in Great Britain by
Clays Ltd, Elcograf S.p.A.

Papers used by Abacus are from well-managed forests
and other responsible sources.

Abacus
An imprint of
Little, Brown Book Group
Carmelite House
50 Victoria Embankment
London EC4Y 0DZ

An Hachette UK Company
www.hachette.co.uk

www.littlebrown.co.uk

To David, Alice and Dominic

AUTHOR'S NOTE

In this book we have used the generally more familiar abbreviations AD (anno Domini) and BC (before Christ) as qualifiers for some dates rather than the equivalent secular terms CE (Common Era) or BCE (Before Common Era) more prevalent in the scientific literature.

Contents

Down, down, down. Would the fall NEVER come to an end? 'I wonder how many miles I've fallen by this time?' she said aloud. 'I must be getting somewhere near the centre of the earth.'

Alice's Adventures in Wonderland by LEWIS CARROLL, originally entitled *Alice's Adventures Under Ground*

[W]hat experiment can be undertaken perfectly free from inconvenience, and all fear of danger, on mountains which vomit forth fire? I would certainly advise the philosopher who wishes always to make his observations entirely at his ease, and without risk, never to visit volcanoes.

The ABBÉ LAZARRO SPALLANZANI, Professor-Royal of Natural History in the University of Pavia, 1798 – taken from *Mind over Magma*, Davis A. Young, 2003

We are such stuff
As dreams are made on, and our little life
Is rounded with a sleep.

WILLIAM SHAKESPEARE, *The Tempest*

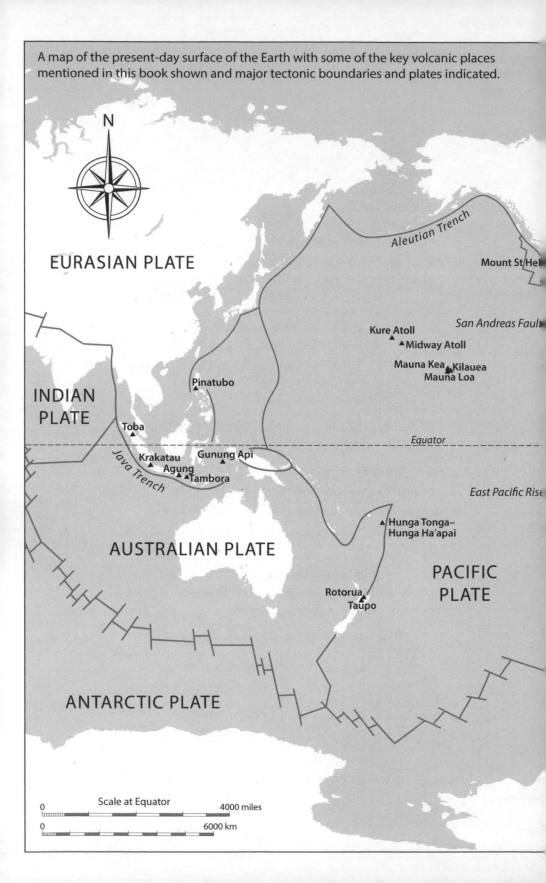

A map of the present-day surface of the Earth with some of the key volcanic places mentioned in this book shown and major tectonic boundaries and plates indicated.

N

EURASIAN PLATE

Aleutian Trench

Mount St Hel

San Andreas Faul

Kure Atoll
▲ Midway Atoll

Mauna Kea ▲ Kilauea
Mauna Loa

Pinatubo
▲

INDIAN
PLATE

Toba
▲

Equator

Java Trench

Krakatau ▲ Gunung Api
▲ Agung
▲ Tambora

East Pacific Rise

AUSTRALIAN PLATE

▲ Hunga Tonga–
Hunga Ha'apai

PACIFIC
PLATE

Rotorua ▲
Taupo

ANTARCTIC PLATE

Scale at Equator
0   4000 miles

0   6000 km

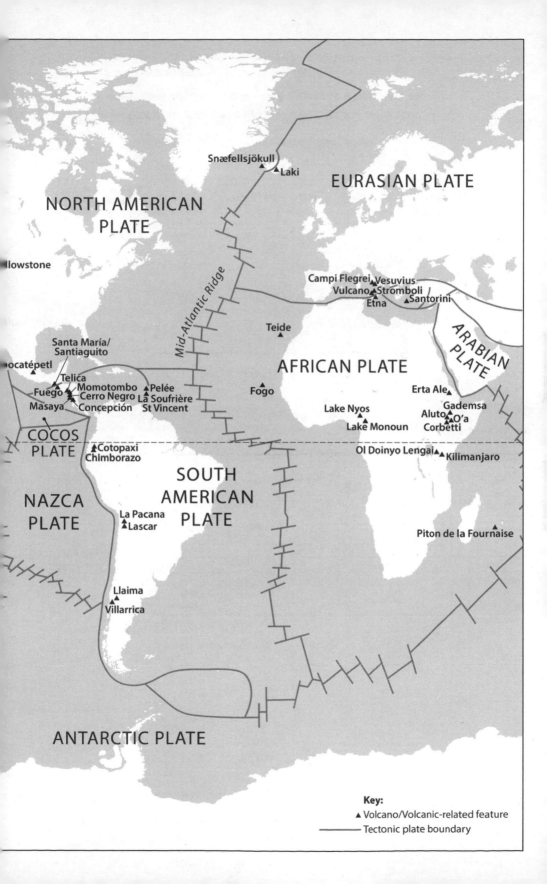

An illustration of Earth's geologic time scale with some of the major trends and events in life's evolution indicated

(MYA = millions of years before present)

Eon	Era	Period	Epoch	MYA	Life Forms	'Big five' mass extinction events
Phanerozoic	Cenozoic	Quaternary	Holocene	0.01	Age of Mammals: Extinction of large mammals and birds; Modern humans	
			Pleistocene	2.6		
		Neogene	Pliocene	5.3	Spread of grassy ecosystems	
			Miocene	23.0		
		Paleogene	Oligocene	33.9		
			Eocene	56		
			Paleocene	66	Early primates	End-Cretaceous, 66 million years ago: wiped out the non-avian dinosaurs
	Mesozoic	Cretaceous		145	Age of Reptiles: Placental mammals; Early flowering plants; Dinosaurs diverse and abundant	
		Jurassic		201		End-Triassic, 201 million years ago
		Triassic		252	First dinosaurs; first mammals; Flying reptiles	
	Paleozoic	Permian		299	Age of Amphibians / Fishes: Coal-forming swamps; Sharks abundant; First reptiles; First amphibians; First forests (evergreen); Land plants	End-Permian, 252 million years ago: estimates suggest over 80 per cent of marine species lost
		Carboniferous		359		Late Devonian, 372 million years ago
		Devonian		419		
		Silurian		444		
		Ordovician		485	Marine Invertebrates: Primitive fish; Trilobite maximum; Rise of corals; Early shelled organisms	Ordovician-Silurian, 444 million years ago
		Cambrian		539		
	Proterozoic	Precambrian			Complex multicelled organisms; Simple multicelled organisms	
	Archean			2500		
				4000	Early bacteria and algae (stromatolites); Origin of life	
	Hadean			4600	Formation of the Earth	

Introduction

The majestic peak of Mount Vesuvius presides over the bustle of the present-day bay of Naples in southern Italy. Houses fringe its lower flanks and at night the mountain's dark silhouette seems to emerge like an island from the sea of the twinkling city lights. The wider metropolitan area of Naples is among the most populous in Italy and the city's story stretches back over more than 2,500 years. Over these long millennia of occupation the region has attracted countless visitors, who flock to enjoy the beauty of its landscape, to savour its food and wine, and, in turn, to marvel at its many layers of human history. The proximity of volcano and people for this prolonged period, as well as the nature of Vesuvius's activity, make this a location of great importance in volcanology as well.

Many key historical observations of volcanic activity and associated phenomena such as gas emissions and ground movements have been made here and it is home to the longest-established volcano observatory in the world – Osservatorio Vesuviano – which has been keeping watch over the area's volcanoes since 1841. The original observatory building can still be seen just off the road, about two kilometres from Vesuvius's summit, where it now houses a museum of past scientific instruments used to study the volcano. The science of volcano monitoring has moved on and these days the area is gridded by high-tech devices sensing the Earth's tremors and changing shape which relay their signals back to large computer banks in the Italian National Institute of Geophysics and Volcanology, Naples in the heart of the city itself – watching and waiting for the next signs of unrest.

Vesuvius was also the first volcano that I personally encountered, during a family holiday to the Amalfi coast in 1987. At the tender age of ten I had no idea of its long significance for volcano science. Nor did I have any inkling of how important volcanoes would come to be in my own life. In fact, my recollection of this first sortie into volcanic terrain is rather hazy. I should perhaps not admit that of my many memories of that holiday it is the hotel's outdoor pool, which, thanks to the relatively cool October temperatures, my sister and I had largely to ourselves, that is probably my clearest recollection through the mists of time, despite the many more profound experiences of that week. But fragments of memory, old photos and a dog-eared holiday diary do recall our visit to the volcano's summit: a short plod from the car park, a large, deep, sheer-sided crater, wafts of 'smoke' clinging to its walls, layers of different-coloured lava and a nervousness about the mountain's potentially destructive nature.

Although Vesuvius has been quiet since the 1940s, to give credit to my younger self, this anxiety was wholly logical based on the facts before me. Prior to our ascent we had visited Pompeii and Herculaneum, both overwhelmed by the infamous AD 79 eruption of the very mountain upon which we stood. The visits to these ruins had marked a vivid impression on me – more so, really, than the dusty trek up the mountain. The death casts of Vesuvius's victims – created by injecting plaster of Paris into the voids left in the volcanic debris once their remains had decayed away – on display at Pompeii made their experiences painfully corporeal: hands often clasped to their faces as they attempted to flee, or curled up in a vain reflex to protect themselves against

The author with her mother and sister at the crater rim of Vesuvius in 1987

the searing inundation. The grandeur and sophistication of these societies were wiped away in just a couple of days. What remained imprinted on my mind was the destructive power of the now inert, and seemingly immutable rock. Rock that had once erupted with sufficient fervour to engulf these places. But, above all else, it was the fear and distress twisted into the bodies of the people it claimed that stayed with me following this first brief brush with Volcanoland.

The AD 79 eruption of Vesuvius was by no means unprecedented but nonetheless it took the local population largely by surprise. Just over a century earlier, the Greek thinker Strabo had noted the similarities between the rocks of Vesuvius and those erupting from Mount Etna in Sicily. But Vesuvius had been inactive for centuries prior to AD 79 and so was not considered a threat. There is evidence that this catastrophe was not unheralded, however. A large earthquake hit the region in AD 62, causing significant damage to local towns and killing hundreds of sheep in the district around Pompeii. These livestock were poisoned, according to the account of the Roman philosopher Seneca, by breathing atmosphere 'tainted by the poison of the internal fires', possibly, as we know now, volcanic carbon dioxide seeping to the surface from underground magma. Minor tremors continued to shake the region after this, but the locals became accustomed to them. In the days before the eruption these tremors increased in strength and were accompanied by other ominous portents, such as springs and wells running dry. Within the mountain the magma was rising.

Today the association between these symptoms and an impending eruption is far better understood, but this is knowledge that has been honed over centuries of data collection and rock analysis. Much understanding has, poignantly perhaps, been gleaned from Vesuvius itself. One of my favourite tales of experimentation with new methods of volcano monitoring at Vesuvius is that of Frank Perret, an American engineer who travelled to Italy for his health and became enraptured by the mountain. In the days prior to Vesuvius's 1906 eruption he found that he could 'hear' a continuous buzzing from below when he set his upper teeth against his iron bedstead. Perret had transformed his own skull into a crude seismometer, detecting the hum created by the movement of magma and gas under the volcano.

These days, well-monitored volcanoes like Vesuvius are webbed by networks of seismic stations designed to capture the first stirrings that signal some change in their inner state. The rock remnants of past eruptions are pored over for clues about future activity. Changes in gas emissions, ground shape, water flow and temperature are all logged as part of a continuous search for hints of impending trouble. But back in AD 79, the people in Pompeii and Herculaneum were not equipped to anticipate the catastrophe that was about to occur.

Could the agonising deaths at Pompeii, recorded in the tortured sinews of the plaster-of-Paris death casts, have been avoided with modern knowledge and instrumentation? As volcanologists we like to hope so, despite the very considerable challenges of getting the right messages to those in charge at the right moment. Although there have been heart-breaking volcanic tragedies in recent times, we might hope that with modern monitoring and the catalogues of past eruptions and their precursors now archived we would have diagnosed volcanic unrest, raised alarms and taken action before magma broke forth. But in the Bay of Naples in AD 79, normal life went on right up to the last moment, when the volcano could no longer be ignored.

At an exhibition of artefacts from Pompeii at the Ashmolean Museum in Oxford in late 2019, I was captivated by the charred remains of a circular loaf of bread found in an oven in the ruins of the city. The idea of a Roman baker putting bread in to bake, completely unaware that almost 2,000 years later it would end up an exhibition centrepiece, reminded me how the day of the eruption began with a veneer of total normality. As the citizens of Pompeii awoke that morning destructive forces were building in the underworld and their daily routines were destined to be shattered by the disjunction of disaster and their relics ultimately transposed into history. But first thing that morning their worries and concerns were those of any other day.

For most eruptions as deep in the past as AD 79, volcanologists must rely solely on the rocks and other geological evidence left behind to piece together the drama that unfolded. But for this catastrophe, there is a human voice to help us interpret the geology, in the form of two letters written some years later by Pliny the Younger to the Roman historian Tacitus. Pliny recounts events from his vantage point across the bay in Misenum, where he noticed that a strange cloud had appeared above the volcano. It grew rapidly, 'more like an umbrella pine than any other tree, because it rose high up in a kind of trunk and then divided into branches'. We know from more recent studies of variation in the size of the rocky chunks strewn different distances from the volcano and preserved in the land-scape that this tree-like column soared 15–25 kilometres above the mountain. Pliny's evocative description echoes through the ages and his name has entered the scientific lexicon to describe this characteristic shape of explosive eruption plumes, so-called Plinian columns. Pliny's uncle, 'Pliny the Elder', set off across the bay on a rescue mission, only to die the next morning chok-ing on volcanic fumes (although a heart attack might in equal part be to blame) having attempted to protect himself from the eruption by tying a pillow around his head as he and his com-panions tried to escape.

Meanwhile, in Pompeii, downwind of the volcano's vent, the town's troubles started shortly after the eruption began, when light-coloured pumice began to rain down, mantling everything like a dusty and abrasive snowfall. The plume would have blot-ted out the afternoon light, turning day to night as if the gods were at work, and in the spreading darkness the streets would have become clogged by drifts of rock as roofs began to col-lapse under the weight of falling debris. No accounts from the city's survivors are preserved but to live through this must have been terrifying: ash would have invaded their nostrils, eyes and ears; pumice would have scoured their skin as they attempted to wade through it; the air would have been tinged with the strange rotten or rasping metallic tastes of volcanic gas. Some likely tried to flee and others thought it best to sit it out in their homes.

Incredibly unpleasant as it was, it is likely that most of those inhabitants who remained would have survived this first onslaught. Death came early in the morning of the eruption's second day as incandescently hot, ground-hugging clouds of pumice, dust and gas – pyroclastic surges – collapsed down-wards from the trunk of the eruption column. Similar scorching avalanches had already engulfed Herculaneum in a matter of minutes during the night. They raced at speeds of up to

100 kilometres per hour burying humans, animals, livestock, belongings and buildings alike. Victims were asphyxiated or flash-heated to death.

One of the first volcanology books I read early in my PhD included images of some of the archaeological remains from Pompeii, pulling the science off the page and drawing me back to my childhood visit. One picture was of the cast of a pet dog, abandoned, chained to a post. Having somehow remained on top of the accumulating pumice fallout, it stood no chance against the pyroclastic surges and its last moments are recorded in its twisted body, arched back and flailing limbs. At 24 years old, grappling with this new science, this image drew me back to the souvenir that ten-year-old me chose to bring home from that first trip to Pompeii: a pottery tile copy of the famous *cave canem* (beware of the dog) floor mosaic excavated from the rubble. I still have it today by our kitchen sink. Like the charred loaf of bread, it is another reminder of the juxtaposition of the mundane and the terrible conjured by catastrophes like that at Pompeii.

While the devastation unfolded in Pompeii and Herculaneum, in Misenum, 30 kilometres across the Bay of Naples, Pliny the Younger had endured thick ash fall, the panic of a fear-stricken mob and total darkness as he waited for his uncle to return. After many hours the dust settled, and a sickly sun, 'yellowish as it is during an eclipse', struggled through the haze. The immediate nightmare was over, but he gazed across the bay to see that a carpet of grey volcanic debris cloaked the landscape where fields, woods, vineyards and towns had once been. The smooth cone of the mountain, as depicted in murals preserved in some of the buried villas, was likely reduced to a stump of itself. The bustling city of Pompeii was obliterated, buried in almost three metres of fresh volcanic rock. The desolation was such that the survivors and onlookers decided the town was beyond hope of redemption. It was abandoned and gradually, over time, the city slipped off the map and was forgotten. Its rediscovery, during the development of new settlements in the late sixteenth century, was coincidentally followed by another violent eruption of Vesuvius in 1631. This eruption was again deadly and destructive, killing many thousands, displacing tens of thousands more and, reportedly, liquifying the dried relic blood of San Gennaro, the Catholic patron saint of Naples.

Our understanding of the Earth and volcanoes has come a long way since the time of Pliny. In part due to Vesuvius's accessibility to Western thinkers (described in 1819 by eminent British chemist Humphry Davy as 'not so small as to be contemptible not so great as to be unapproachable') and long popularity as a tourist destination, the study of its spate of eruptions between 1631 and 1944 played a significant role in these advancements. Ancient and medieval ideas of the character and causes of volcanism – of deities, as chimneys for the Earth's inner fire or as gateways to hell – have developed with the intervening centuries of observations, data and science and an ever more sophisticated global perspective. Our studies are often driven by the human experience of volcanism. How do we map volcanic hazards? How do we avoid human catastrophe and tragedy like that which unfolded in the Bay of Naples? How does an active volcano change its environment and how does this affect local people? How can it be that we feel the effects of a large volcanic eruption on the other side of the planet? Volcanoes are often, sometimes out of urgent necessity, defined by the characters and concerns that we project upon them. But when peering into the mouth of a volcano, the dominant feeling is often one of lingering insignificance. In the face of this overt expression of our planet's power, human meanings and our apparent agency can seem to recede and something else, perhaps more fundamental, comes into view.

The lifespans of volcanoes inhabit a timescale far different from our brief geologically ephemeral, little human lives. The

An explosive eruption of Vesuvius viewed from Naples in October 1822 and published in Scrope's Considerations on Volcanoes *(1862). Note the shape of the volcanic ash plume as like an umbrella pine tree, as described by Pliny, and the volcanic lightning and curtains of debris falling from the plume.*

activity of Vesuvius punctuates European history: from Roman times to its most recent eruption, in 1944, just after Allied forces took control of Naples during the Second World War. However, the few-thousand-year span of our recorded history is a mere fraction of the mountain's age. The oldest rocks building Vesuvius's bulk date back far further, to about 300,000 years ago, roughly the same age as the oldest known remains of our species found in Africa. Some volcanoes are much younger, of course. Parícutin in Mexico, for example, famously emerged from a farmer's field one otherwise normal February day in 1943, building its cone to over 100 metres by the end of its first week. But when we gaze upon the majesty of Vesuvius it is a view that captures not just ourselves in the awe of the vista but, by some measures, the timescale of humanity itself.

Even this mind-blowing time slice charting hundreds of thousands of years is only a blink of an eye for a planet 4.5 billion years old. If we shift our gaze from Vesuvius and take a broader view of the rocks that make up our planet's surface, evidence for volcanic activity in the form of dated lavas goes back billions of years. Volcanism of some sort has been part of the rocky ball we call home since its very beginnings. Although individual volcanoes come and go, volcanism as a process has persisted across the mind-bending eons of Earth's history. Magmatism has been a part of our planet's intrinsic character in a way that we simply are not. Volcanoes were here before us and they will be here long after. The guard may change, but they are sentries of our planet's deep and ongoing history.

Building a habitable planet is a complex process but it is no exaggeration to say that volcanism has been fundamental to shaping this beautiful world we call home. The heat sources in the early Earth and the continued bombardment by asteroids and other planetesimals would have led to it being substantially, or perhaps even entirely, molten at times during its first 100 million years. Some evidence even suggests that it could have been covered in large, maybe planetary-scale, magma oceans during this formative period. This was a time when volcanism, or perhaps it is better to use the more general term magmatism, was the planetary rule rather than anything exceptional. This world of black and orange broiling expanses is almost unimaginable, the stuff of science fiction, but written into it somehow were the blueprints for our present-day globe; our gloriously habitable blue marble.

Although it might seem contradictory, volcanism played its role in transforming Earth from a lava world to a water world and has since helped to keep it a blue rather than a black and red planet. It is not known precisely when or how the atmosphere formed and there is still much to understand about its evolution over geological time. Nonetheless, the release of volatile molecules like water and carbon dioxide from the molten Earth was, and continues to be, undoubtedly very important. As the heat of the nascent planet cooled, the water vapour in its gassy envelope condensed and fell as rain, flooding the Earth's surface and forming the primeval oceans. Magmatism and volcanism were again the engines that helped to forge the first land: fashioning islands and then continents, pushing this new land out of the seas. Those early primitive lavas would have broken the waves in a cauldron of steam, gradually building into gently sloping shield-shaped volcanoes. The emergence of these primordial landscapes was witnessed by a dim, young Sun and a giant Moon, much closer to Earth and about double the size of the disc we see in the night sky today.

Volcanoes were there when the first basic life evolved somewhere in our planet's ancient waters. It may even have been volcanic heat, in deep-sea hydrothermal vents, or volcanic lightning during explosive eruptions, that helped to rearrange some of Earth's atoms into the first primitive molecular building blocks, allowing biology somehow to begin. Much later, volcanoes looked on as plants spread to cover the continents, and the first animals crawled from the sea to make the land their home. As geological time has ticked on, the rocks have witnessed the planet's story. Species have waxed and waned, sometimes gently but sometimes abruptly in the great dyings of mass extinctions, when immense swathes of species have been cut out of history in a geological flash. Volcanoes likely played a part in these events, too, although many questions and much mystery remains about their exact role.

The most compelling component of Earth's story will always be that of its abundant and varied life, culminating in the richness we see around us today. For this world to be possible the fluctuations in its environment have needed modulation. However, it may sometimes seem, Earth is not, in fact, a place of extremes relative to our interplanetary neighbours. We do not need to cast our eyes far into our solar system to realise this. The temperature difference between lunar night and lunar day is about 300°C. Our sister planet Venus, so alike in size and make-up, has average surface temperatures 450°C hotter than here. Our smaller neighbour Mars is 80°C colder on average. Agents like active volcanism, while holding the potential to sow great change, have also helped to define the limits of Earth's environ-mental variability. Volcanoes are a part of a system of terrestrial checks and balances which have allowed life to persist, but which we still don't fully understand. Venus and Mars show us how differently things could have been. There are many important factors governing a planet's evolutionary path but having just the right mix of tectonism and magmatism wrought into Earth's planetary cycles has been part of its journey to habitability. These so-called mountains of fire are a sign that our planet is geolog-ically alive, and their constant maintenance of its atmosphere, oceans and continents has allowed the conditions for evolution to chart a path to our existence as a species. Volcanoes underpin our history, even when we are not directly aware of them. They are an integral part of the rich tapestry of the world upon which we find ourselves, where we are able to marvel at its bounty and beauty and to try to reason out its meaning.

I came to volcanology later than many of my colleagues – sev-eral of whom have been possessed by these powerful mountains since childhood – and somewhat by accident. The title for this book was, in part, inspired by Lewis Carroll's sometimes per-plexing *Alice's Adventures in Wonderland*, a childhood favourite. As in Wonderland, things in Volcanoland are often not quite what they might at first seem, and this is part of what makes it so fascinating. Even more than regular mountains, volcanoes represent great danger. They can lash out to wreak havoc and human despair. It is often the reason they make the news and, when I started my PhD in volcanology, this is how I quite understandably saw volcanoes too. But when you travel to volcanoes and talk to those who live on and near their flanks, a much more complex relationship between humans and volcanoes quickly emerges.

Volcanoes can devastate and destroy but they are also revered and can be integral to local lives and livelihoods: fertilising soils, attracting tourists or providing power, water or raw materials. Sometimes people live near volcanoes because they have no choice, but sometimes, too, they choose to live there because of the opportunities that they provide or because it is the place they hold as home.

Through my studies I have come to think about some volcanoes as individual characters. I am not alone in this. These rumbling mountains can easily seem alive and often take on powerful personas, especially for those living with their daily presence. An active volcano combines both the bulk and immutability we might feel gazing at other mountains with something also dynamic, shifting and at times capricious, and thus in some ways far more human. Their landscapes can change as fast as our lives. In a six-year gap between expeditions up Etna due to the birth of my two children, a new cone had built up on the volcano's south-east summit, and at its central peak, the ridge between two of its craters had all but disappeared, having slumped back into the abyss. Seeing it again in 2012, it felt almost as transformed as my life and I was caught by something that felt a little like empathy. For Etna though, this was nothing so profound and just another tiny increment in its restless but nonetheless fundamentally inanimate ebb and flow. We gaze upon volcanoes and they amaze us, but however much we deify and anthropomorphise them, however much we push and prod them, they have absolutely no opinion of us.

My fascination with volcanoes is very personal at times, but it is also deeply scientific. As much as we might sense them as alive and full of individual character, this is just a trick of the mind; our best dialogue to understand them comes through careful observations and the documentation of commonalities, patterns and trends. Volcanology has drawn on all the tools laid out not only by the great expositions of geology but also the fundamental findings across the breadth of human sciences. Ours is a science woven from these strands over the centuries of human endeavour into a unifying theory of how, why and where we get volcanoes on our planet. Our understanding of the substance of volcanism on the molecular scale gives us a human-scale perspective on why some eruptions are more dangerous and deadlier than others.

Like the structures of these most dynamic of mountains, human knowledge of their nature has developed in layers over the millennia. Edifices have been constructed that have been cut back by the ruptures of new ideas that then build new cones of thinking to greater heights in their place. We undoubtedly now know far more than we ever have before, but if the history of science teaches us anything it is that some ideas that we now hold dear will turn out to be at least incomplete, if not entirely wrong. We inhabit a brief moment in the lifespan of our favourite volcanoes and a specific moment, too, in the journey of human thinking. From this vantage point, I have spent the last 20+ years trying to understand how volcanoes erupt and their different eruptive styles, how we might predict their behaviour, how humans experience them and their impact on our environment. They have the ability, too, to stretch beyond their local surroundings to change our planet's wider environment and this challenges us to revel in the richness of thinking of them not only as individual entities but collectively as a planetary-scale process that shapes and reshapes our and other worlds.

In this book we will journey through some of what we know about volcanoes, how we came to this knowledge and where we are going next. Each chapter poses a question that I seek, at least in part, to answer. It is an expedition that starts within the Earth to explain how and why volcanoes occur where they do, how they erupt and why one can erupt so differently from another. Much of our knowledge of volcanic behaviour comes from reading the rocks, and we will explore how we move through a volcanic landscape to reconstruct past eruptions from the debris left behind. But volcanoes reach beyond themselves as well, stretching out to change or maintain the local, regional or global environment. This is largely through the gases that they emit, and to understand their wider influence we must quantify what gases these are and how and where they are released. We must also push to observe and model the effects of these emanations in the short- and long-term balance of the atmosphere.

These tools arm us to see volcanoes both as they are in themselves and also collectively as a world-shaping process, and this understanding yields new insights into the subsurface machinations of our own world and also beckons us to look out to the heavens. For what we understand of Earth's volcanoes, illuminates the role they have played in shaping other planets too, especially as new data arrives from ever more detailed missions to our celestial neighbours and ever more sophisticated observations of exoplanets beyond our own solar system. This in turn tests our understanding of our planet and its history. But it is not just other worlds that can be foreign and remote to us. Our planet's own past contains evidence of volcanic events or protracted episodes of magmatism operating on a much larger scale from those in the brief history of human observations. Understanding volcanoes today is a key part of comprehending the impacts of this past volcanism and how it changed the course of Earth's biological evolution.

There are many routes through Volcanoland and many

different perspectives, far too many for me to capture fully either scientifically or geographically within these pages. My account is, of course, biased by my scientific background and interests and the places that I have been lucky enough to visit. Having been neither born nor raised in a country with active volcanism I am always a visitor in volcanic locales but I hope to give some sense of what it is like to live and work in such places and how these powerful landforms and their activity have captivated people through the ages. I find it fascinating to look at a volcanic landscape furnished with the accounts of how others have seen it before me. It is part of sensing the timescale of these places and imagining the spans of time outside ourselves inhabited by previous generations of human observers and the far deeper time still of Earth's geology stretching behind and in front of us. There is a profound and growing awareness of the planet's precarious future in the augurs at the moment, and perhaps volcanoes can help us in some way to make sense of this too. My path through Volcanoland has also involved a shift in the personal insights that I have gained from studying magmatism. I started out mainly motivated to understand volcanism in terms of its power to change our planet's past, present and imminent future. More recently, I see, especially in the very greatest episodes of Earth's past volcanic activity, a lens through which to view lessons about humanity's potential as a geological force and the responsibilities this entails.

Working in Volcanoland is not always easy. Some days I despair that, for all our achievements, our science cannot yet always avert the loss of lives or livelihoods. Some days I return from the field dusty, bruised and battered, other days none of the data makes sense. But, to my mind, journeying into Volcanoland is one of the most compelling adventures we humans can go on.

Part One

The Rocks of Volcanoland and How They Erupt

Chapter 1

The Molten Earth

What causes Earth's volcanoes?

Dwelling as we do on the surface of our world, we tend to consider rock to be solid, the stuff of cliffs, mountains, boulders and building stone. But within the depths of the rough sphere of rock we call home, this fundamental fabric can behave very differently. At the great temperatures and pressures inside the Earth, solid rock can slowly flow or creep and, in some places, it melts and moves as a liquid or a magma. Sometimes this molten Earth finds its way to meet us at the surface, spewing forth in towering explosions, fiery fountains or flows and creating the structures that we call volcanoes. But on the present-day planet, active volcanoes are not the norm and in many parts of the world in the absence of textbooks, travel accounts and media coverage, you might easily go a lifetime unaware of their existence. A primary question for any adventure in Volcanoland must be what are the underlying causes of volcanoes? Where does the molten Earth – the magma – come from and why are volcanoes where they are? This is a journey that truly started for me when I began my PhD in 2001, but the quest to understand the origins of volcanism is a voyage of human knowledge casting back millennia.

I turned 25 during my first research expedition to an active volcano at the end of the first term of my PhD. I was working on Masaya in the heart of Nicaragua, one of the Central American nations forming the delicate isthmus that bridges between the great hulks of North and South America. Masaya is not a volcano that figures especially prominently either in the landscape or the history of volcano science. Nonetheless, it was the beginning of my apprenticeship in volcanology, and the first place I got to peer into a fuming volcanic vent and try to make scientific sense of what I was seeing. Although there were many precursors, Masaya is where I feel that my adventures in Volcanoland properly began.

The author running gas and aerosol sampling equipment on Masaya's crater rim, 2001

That December we set off early in the cold and dark from Cambridge towards Heathrow airport, with four bags straining full of equipment. It was dark when we arrived in the Nicaraguan capital Managua, too, delayed and three bags short. Even in the dead of night it felt hot and humid. I had not slept on the journey, in excitement, and I was a little delirious. The pale moon hung over Lake Managua, as we grappled with missed turns to find the hotel. It was my first time in Latin America and I vividly remember catching snatches of the railing-fronted breeze-block houses and the illuminated cola-sponsored bar signs in the villages we passed. But it was too dark to see anything much of the surrounding landscape and when we finally arrived at the hotel and crashed gratefully to rest I had no real feel of where I was or any sense of my proximity to an active volcano.

The next morning, I woke first and stumbled out onto the porch with the combined dizziness of jet lag and hunger. Pineapple plants, a lawn of thick-bladed grass and strange tropical trees and creepers greeted me. And right there, looming not so very large above the lush vegetation, was the volcano itself. In the relative cool of the morning, a cloud was nestled around its top and the barely perceivable shapes of birds soared against the billowing white fumes. I could just make out a wooden cross near the rough summit and scrubby bushes, trees and smudged grass rose almost all the way up to this symbolic marker. There was a rusty scar to the right where the lobes of a rubbly lava flow had reached over the crater rim, frozen in the act of tumbling.

As I mentioned, Masaya is a rather understated volcano, at least from afar, with its highest point only a touch over 630 metres above sea level. One of Nicaragua's main roads, Route 4, runs along its north-eastern side but driving from the town of Masaya to Managua you would be forgiven for missing it as you passed. Its topography is subdued and complex, something of a footnote on the landscape compared to its majestic cone-shaped neighbour Momotombo, which dominates the northern shore of Lake Managua nearby. There is a view of Momotombo that is suddenly revealed as Route 4 crests a hill, its elegant cone reaching for the sky like steepled fingers appearing as if from nowhere, that takes my breath away every time I see it. It is only an illusion, but you could almost believe that the main highway is leading you to the very foot of the volcano. Masaya, in contrast, lies within a large pinched oval shaped depression – a caldera – steep rimmed on three sides, almost as if hinged along its north-eastern edge, with Lake Masaya nestled foetal-like against its eastern cliffs. The caldera is a national park and a tarmac road snakes its way right up to the volcano's low summit.

What it lacks in height Masaya makes up for in sheer volcanic busyness. Only 30 years after Christopher Columbus landed in the Americas, the first *conquistadors* arrived in Nicaragua. The volcanoes were not shy, and the colonists were greeted by eruptions of Masaya and Momotombo in 1524. In a letter to the Spanish emperor relaying the conquest of Nicaragua, the local governor Pedrarias Dávila details the first volcanic activity observed by the conquistadors in these new lands: 'In this province of Masaya there is a large mouth of fire which never ceases to burn and during the night it is so big as if it reaches the sky ... there is light as if it was day ...' Human history might have moved on but Masaya and Momotombo continue to 'burn' on and off five centuries later.

Masaya today is a popular destination for tourists and school trips. At the crater rim there is a car park, often full of yellow US school buses, imported to Central America to see out their retirements as local transportation. Many of them have been repainted in vibrant new colours: fiery reds, iridescent blues, emerald greens and deep purples. Their exhausts periodically belch out sooty fumes and the buses often struggle up the hills, but their horns – which are deafening if you are close by – have lost none of their potency. Just a few steps from the parking spots, people can peer into Masaya's active Santiago crater and watch noxious hazy gas billow out of the Earth's insides.

The crater is a moonscape. Its cliffs are sheer and striped with interleaving seams of frozen lavas and layers of volcanic ash or cinder. The effect is like a slice through a red, ochre and grey layered cake; layer upon layer of volcanic process, Masaya's history laid bare. The crater floor, about 150 metres below, is an arena of twisted lava, dust and rubble. When I first gazed into the volcano, I could see the word 'Santiago' spelt out in white rocks arranged by some intrepid adventurers who had braved the descent. On my second trip, in 2003, I watched as colleagues abseiled into the crater and dispersed the letters at the request of the national park chief. Scalloped out of this terrace and further down still, are the active vents, very much like mouths, where the red-hot inner Earth meets the surface and the air. In 2001 there were two vents and, but for the billowing fumes and occasional clatter of small rock falls, superficially all looked relatively calm. Whenever I visit Masaya this most intimate part of the volcano has changed, as the volcano breathes and shudders, and new details are added to its story.

If you are working on the volcano you can get permission to drive through the barrier blocking the way beyond the tourist area. A road skirts along the crater rim to an abandoned car park on the opposite side. The ground here is fissured from volcanic wear and tear, covered with a dark dust and strewn with stones and boulders of varying sizes. When it is windy it becomes quickly apparent why it has been deserted. Once the hot gases loft out of the depths, the prevailing winds, blowing across the crater's top, pick them up and push them over the rim. In the heat of the day these fumes are often transparent and invisible, but they have other ways of making themselves known. Even when there is nothing to see but the hint of a ghostly haze, the plume can rapidly fill your lungs and have you choking and clutching for your gas mask. Although still often described outside volcanology as smoke, this corrosive cocktail's chemistry bears little resemblance to a face full of billows from a bonfire or second-hand whiffs from a cigarette. These fumes are full of rasping acidic gases and – fresh from the Earth's innards – they are exactly what draw me and many other volcanologists to Masaya. The volcano is a natural laboratory where we can take measurements and run experiments and try to determine what is going on inside the crater and, indeed, the Earth. These pungent vapours, however superficially unpleasant, are catnip for gas geochemists like me.

Despite the fumes and discomfort of working in a rubber gas mask, this 'gassy' car park is a special place. There is very little vegetation but some surprising wildlife. In a certain spot bees burrow into the flaky compacted volcanic dust and mud, digging cell-like nests, shielded from predators and parasites by the inhospitable fumes. When the wind drops enough you can hear them buzzing. You also get a different view of the active crater from here. Looking across towards the tourist car park, the stratigraphy – the geological layers – of the older Nindiri crater are revealed where it is cut by Santiago's newer activity. The layers sag between the volcano's sides, like an elaborate igneous version of the hammocks on sale in Masaya artisan market. The cross that I had spotted that first morning presides from a hill on the far side of the crater. The original was apparently erected by a zealous Spanish friar, Francisco de Bobadilla, in the 1500s to exorcise the demons that might have escaped from the volcano's mouth, believing it to be a back door to hell. Dark vultures with white tipped wings circle overhead, riding the thermals created by the volcano and the Sun. Occasionally they perch on this cross, and I imagine that they watch us from afar as we work. They are beautiful but ominous, a reminder perhaps that there is always an element of danger, being on an active volcano. While Masaya's gases are generally the biggest hazard, the volcano can also lash out more violently. Just the April before our 2001 visit, with little warning, an explosion opened a new vent in the crater floor, throwing out boulders (volcanic bombs) and ash, setting grassland on fire, damaging cars and meting out injuries (thankfully in this case only minor) to the visitors present: a reminder of the brooding power behind one of Nicaragua's foremost tourist attractions.

One of my favourite times of day here is sunset. Often the wind seems to drop and the sky morphs into pinks and oranges as the Sun dips towards the unseen Pacific. A colony of parakeets lives around the crater and if you are lucky you will see the glint of their green tails and hear their calls as they return home. As the Sun continues to descend, the colours wane from the world around and the first stars start to creep out from the heavens above. Below, within the crater, the pigments from the sky almost seem to drain into a new glow emanating from the active vent. The hot stuff. The magma. The molten Earth.

Back in 2001 this night-time glow was very subtle. Nonetheless it was fascinating, like staring into the last embers of a campfire. In quieter moments I could hear the roar of the magma as it raged below the ground out of view. The dim light would pulse just perceivably, as gas slugs burst from its surface and sometimes, in the cooler evening, I could feel the hint of the warm gas wafting up and enveloping me like a blanket. Moments like these on a volcano feel timeless. But, of course, they are not. Volcanology papers serialise sketches recording volcanic places in great detail; like scientific flipbooks telling the story of how these restless features change. Vents open and close, the magma levels rise and fall, fiery spatter hardens into new rock. The two vents I saw on my first trip to Masaya were just another page in the sequence, and every time I return I wonder how this inner sanctum of the volcano will have altered. Sometimes the changes are subtle: a new vent will have opened, another might have been filled in by fallen rocks, there is a stronger glow at night than there had been previously, or perhaps this glow has dimmed. Other times the changes are much more dramatic. On a later visit, in 2017, Masaya's active vents had grown and coalesced and the floor had fallen in to reveal the top of the magma-filled conduit within. At night the glow lit up the crater like a roaring autumn bonfire. You could see it as you rounded the final bend on the road up to the summit. More tourists than ever flocked to view it, and the night-time view across the crater from the gassy car park was lit with camera flashes as well as by lava. The inner crater had become a pit of raw ceaseless energy – roiling, red and restless. I was surprised by how fluid, how dynamic, it was, like a window onto a tumultuous fiery sea inside the Earth rather than static, as implied by the technical term 'lava lake', which had been used in the formal reports describing this latest activity at Masaya. It felt like a deeper glimpse into the processes that I had been studying for years without seeing their actual source, like I was being let in on a secret.

Held up to the temporal yardstick of written history in the region, Masaya's activity could appear endless. The rocks tell a different story. Its caldera dates from around 2,500 years ago so Masaya, like all of us, had a beginning, and there are many important events in its history that took place before we have written records to compare with its current state. Nonetheless, the accounts of the volcano chime through the ages in important ways. The indigenous people of Nicaragua worshipped the volcano as a god and tales of human sacrifice and dark sorcery waft around it, as recorded by the conquistadors. There was a lava lake at Masaya when the first chronicler of the 'Indies', Gonzalo Fernández de Oviedo y Valdés, scrambled up in July 1529. Oviedo describes it as 'a fire that was liquid as water and that matter was burning more fiercely than red hot coal and more ardent in colour; burning more than any fire can ever burn, if that is possible'. These words leave little doubt that what Oviedo saw more than 450 years ago was very similar to the tourist spectacle in 2017 giving us some insight into the longevity of Masaya's current volcanic character.

But in the 1500s these ardent colours awoke greed as well as awe in the foreign visitors, with some convinced that there was gold and silver in the lustre of the lake. Friar Blas del Castillo was especially persistent in this belief and in 1538 he began prospecting. His efforts were initially kept secret because he and his colleagues did not want to share the treasure. The friar conveniently justified his secrecy thus: '[K]eep your silence, God doesn't want the gold to be discovered by the rich, but rather by the poor and humble.' After carefully surveying and covertly carrying the necessary equipment to the crater rim, the prospectors were ready. Friar Blas was slowly lowered into the crater inside a basket. Health and safety was not forgotten: he wore an iron helmet hidden under a straw hat to protect his head, and carried a wooden cross to ward off demons, as well as a flask of wine to fight thirst and fear. Unsurprisingly, my risk assessments for field work today look considerably different, with spirits of the alcoholic kind strongly discouraged while on the volcano and no consideration of the hazards presented by spirits of a more supernatural variety.

This first hell-in-a-hand-basket descent was not successful. They made further attempts, but all the prospectors came back with was dark basaltic rock that, after analysis at the Foundry in Leon, revealed neither silver nor gold. The frustrated governor, now aware of these activities, banned Friar Blas, who died soon after, from any more prospecting. But Friar Blas's enthusiasm for volcanic riches was infectious, and others took up the quest, with the schemes getting ever more elaborate. In 1551, the Dean of the Cathedral of Leon made an unsuccessful request to the emperor for 200 slaves to dig a tunnel through the crater's walls in order to empty it of all its gold. But by the time the taxation rate to cover these efforts had been agreed and the requisite permissions given, the volcano's humours had shifted and the lava lake had cooled and frozen. The glitter of gold was gone.

*

It is easy to dismiss these sixteenth-century observers, naively taking all that glisters for gold and magma for fierce burning matter; but getting to the heart of *what* a volcano really is took a long time for science to piece together. Gazing into Masaya's mouth that December day in 2001, I was at the start of my ongoing journey to understand. Given the mysterious heat and power of volcanoes, their association with deities and demons – at Masaya and across the globe – does not strike me as surprising. Explaining the source of this raw power and its ability to melt rocks is part of how we comprehend the inner workings of our planet, how it has come to be, and our place on it.

As a child I was fascinated by flames. I still am, in fact. It is natural to be entranced by them, and I see it in my own children as they grow. Birthday candles and bonfires, a cosy fire in the grate at Christmas, leaning carefully towards it for comfort. The heat from these flames is chemical, released as the atoms in the wax, wood or coal reorganise themselves and tangle up with oxygen in the air to form new molecules. Deprive the fire of oxygen or fuel or douse it with water and it goes out. Whatever visual similarities it might have, though, volcanic heat is something else. Nothing is burning. Nothing can be put out.

The first time I stood up close to a lava flow was a July evening in 2006 on Etna in Sicily. The ground throbbed, the air stung and everything appeared red, orange and black. Above me a pulsating fire fountain roared. Bright trails of debris flew from the volcano's mouth, then cooled and dimmed, creating a cone of broken black around the drama. A stream of lava wound its way out from the base of this cone and flowed down towards me with an unexpected metallic sound, clinking like a chain being dragged along concrete. Gazing into this river of red, I leant towards the flow until I felt the heat get too much. The intensity was shocking and I pulled back, afraid that the margin between safety and severe danger was too thin for me to handle. Unlike the fire in the hearth, safely contained, there was nothing cosy or comforting about this heat.

Our widespread use of fire as a species dates back at least 100,000 years. It is thus perhaps unsurprising that early attempts to make sense of volcanic activity drew upon this primitive, more ubiquitous experience of fire, flames and burning. The coincidence that the cradle of European civilisation in places like Greece and Italy is co-located with a collection of active volcanoes means that we have records of how people from Western antiquity thought about volcanism and its causes. Activity at volcanoes like Etna, Stromboli and Vulcano – named after the Roman god of fire and metalwork, from which we take the term 'volcano' – on and around Sicily and Santorini in the Greek Cyclades, is a line of shared experience between ancient and modern inhabitants. Stromboli has been continuously active as far back as written history goes, more than 2,500 years in that part of the Mediterranean. Its regular explosions still delight, and sometimes terrify, tourists today. At night the bursts of hot rock illuminate the tiny particles within the gas, pushing curtains of light into the sky.

These same great gasps of hot gas led early Greek seafarers to imagine that the giant Aeolus, the keeper of the winds, was imprisoned beneath the islands. Though the myth is long since debunked, Aeolus remains captured in the name of the collection of islands that includes Stromboli and Vulcano. These volcanic 'Aeolian Islands' poke out of the cobalt Tyrrhenian Sea in a complex arcing cluster off the north-east corner of Sicily. Ancient Greeks and Romans explained volcanic activity as caused by fires within the Earth, or great winds heated by friction that were being breathed out. These ideas coalesced and matured, and the Romans imagined that volcanoes were smokestacks for giant internal furnaces fuelled by coal, bitumen or sulphur inside the Earth. They theorised that there was a great network of flues within the planet and that volcanoes were where these inner fires burst forth, like safety valves.

These classical ideas of volcanoes fed into medieval Christian images of hellfire. Volcanoes were the chimneys of hell, their activity a consequence of human sins, their roars the screams of tormented souls. It was these ideas that led Friar Blas del Castillo to include a cross in his field kit on his journey into Masaya's crater and Friar Francisco de Bobadilla to erect the first cross on its summit. Even with the advent of the Renaissance, the concept of Earth's internal fire dimmed only slowly. Among the many treasures in Oxford's Bodleian Library is a copy of *Mundus subterraneus* published in 1664 and written by the prolific German Jesuit scholar Athanasius Kircher.

Fleeing war in his homeland he travelled to southern Italy where he witnessed the volcanism of Etna, Stromboli and Vesuvius, leading him to ponder its nature and origins. Some years ago, one of the illustrations from Mundus subterraneus formed part of a Bodleian exhibition of the volcano-related riches lurking in its scholarly archives. Kircher's cross section through our world was one of my favourite images on display. The engraved Earth is a dark circle rimmed by no fewer than eleven active, conical volcanoes – 'fire-vomiting mountains' as Kircher would later refer to them – evenly spaced around its surface. The volcanoes neatly bookend distinct worldly domains, with six sectors of oceans on which ships bob, and five continents illustrated as mountain ranges. The heavens above are a tumult of cloud and wind. But it is the inside of the Earth that most immediately draws the eye. Kircher had taken note of miners' reports that temperatures increased with depth in the Earth's crust, and he captures this inner heat as a massive blossom of flame in the very heart of the planet. A dendritic network of channels reaches out from this central fire to many smaller flowers of flame, nodes in a complex web that ultimately vent via volcanoes at the surface.

But with the passing of time, there was a change in how volcanoes were envisaged in Western philosophy. Hell itself was challenged, reimagined or even subjected to a form of empirical analysis. In his 1714 book *An Enquiry into the Nature and Place of Hell*, Tobias Swinden, rector of Cuxton in Kent, reasoned by comparison to the heavens, that any earthly subterranean space 'must be too small to contain the lapsed Angels, and the infinite Numbers of the Damned'. For a believer like Swinden, hellfire was real but he moved it from within the space-limited Earth to the Sun. Other thinkers of the time proposed different theories. Isaac Newton's successor as Lucasian Professor of Mathematics at the University of Cambridge, William Whiston, suggested Halley's Comet as a possible location for hell and its interminable fires. Others rejected a literal view of hell altogether as Enlightenment thinkers pushed to separate the material from

An engraving showing a cross section of the Earth and depicting the world's volcanoes being fed by fires connected to a central source inside the planet. From Athanasius Kircher's Mundus subterraneus *(1664).*

the spiritual universe, or rejected religious orthodoxy entirely. While explanations of volcanism drew away from hell's eternal damnation, their association with more benign forms of fire took longer to put out.

Fire remains at the heart of volcano taxonomy even today. Volcano names like Fogo ('fire' in Portuguese) in Cape Verde, Fuego ('fire' in Spanish) in both Guatemala and the Canary Islands and the Piton de la Fournaise (French for 'Peak of the Furnace'), the first volcano I ever saw erupting, on Réunion Island, record the reach of European colonialism. But languages across the globe capture volcanoes as fiery or smoking mountains: from individual volcano names like Erta Ale in Ethiopia ('smoking mountain' in the local language) and Gunung Api in Indonesia ('gunung' for mountain and 'api' for fire, flame, blaze, light or glimmer) to the general words for volcano meaning fire or burning mountain in places as far flung as Iceland (eldfjall), Japan/China ('kazan' or 'huǒ shān') and Samoa (mauga mu). Fire is also branded into the technical language of volcanology. The shattered fragments of rock found after volcanic activity are pyroclasts (Greek for 'fire' and 'broken in pieces'). The finest sand-like products of volcanic explosions are known as 'ash'. Vigorous displays of volcanic activity in places like Kīlauea on Hawaii or Etna are described as 'fire fountains' – just right to convey what is seen. These terms are evocative and help to relate the unusual phenomena of volcanism in a shared lexicon of experience, even if they are technically wrong. As scientific understanding of volcanoes and their products – and of fire itself – changed, the challenges of unravelling what was happening under the surface mounted. It was hard to imagine a sufficient stash of fuel continuously ventilated by oxygen within Earth's depths to sustain its inner heat. And, despite their names, the rocky products of volcanism bear little resemblance to the ash and cinders from a forge or furnace, so in the seventeenth and eighteenth centuries thinkers began to look for other ways to account for volcanic heat.

As chemical knowledge grew, new explanations were developed for volcanoes, beyond simple fires within the Earth. Natural philosophers like the French academician Nicolas Lemery experimented with the 'fermentation' of iron and sulphur as a potential source of heat to ignite Earth's internal fires or to crack open ventilation for air to fuel the flames. In 1690 he recorded experiments in which he buried a paste of wetted sulphur and iron powder beneath the ground. The swelling and heat generated convinced him that this was 'fully sufficient for explaining after what manner fermentations, shocks and conflagrations are excited in the bowels of the Earth, as happens in Vesuvius'.

The discovery of new elements ignited new theories of the origins of volcanism as well. The Cornish chemist Humphry Davy was the first to isolate the raw elements of potassium and sodium. The reactivity of these alkali metals with moisture is impressive. I have a vivid memory from school of my teacher demonstrating the reaction of pure potassium and water. The metal was so reactive that it had to be stored in oil to isolate it from atmospheric oxygen and moisture. Once liberated to the water, we watched the potassium melt into blobs and fizz around the surface of the clear trough accompanied by sparks and lilac flames of burning liberated hydrogen. While today this demonstration is combined with other experiments to illustrate important trends in chemical reactivity, in the 1800s Davy was sufficiently struck by it to draw other parallels and it led him to propose a new theory of volcanism. In 1808 he reasoned that '[i]f the interior of the globe had been from all time in a state of ignition, the effects must have been long ago communicated to the surface, which would have exhibited not a few widely scattered volcanoes, but one glowing and burning mass'. He suggested instead that large subterranean supplies of reactive metals in some places caused the localisation of volcanism and illustrated his theory at places like the Royal Institution in London by

constructing spectacular artificial volcanoes out of clay, packed
with metals like potassium and ignited by water.

Arguments about the source and depth of volcanic heat raged
on along with questions about how exactly the temperature of
the Earth changed during descent within its sphere and whether
the planet's rate of volcanic activity was constant or diminishing
over its history. These theories of Earth's heat fed into the imaginings of literature too. When I was a child we had Jules Verne's
1864 classic *Journey to the Centre of the Earth* on cassette tape
to keep us entertained on long car journeys. This epic tale, in
which the characters decode a runic note that points the way to
a tunnel to the centre of the Earth through Snæfellsjökull volcano in Iceland, enlivened many a monotonous motorway mile.
The correct crater to enter is indicated by the noon shadow of a
nearby mountain peak just before the end of June. Descending
into the Earth's hot interior seemed problematic to me, even as
a child, and one of the characters, Axel, shares concerns about
being overcome by the heat as they journey down. But Verne
hangs the story around Davy's metal-fuelled theory of volcanic
heat. As the eccentric Professor Lidenbrock explains, '[t]he
Earth has been heated by combustion on its surface, that is all.
Its surface was composed of a great number of metals, such as
potassium and sodium, which have the peculiar property of
igniting at the mere contact with air and water.' This convenient
ambiguity about the nature of volcanic heat allows Verne's travellers to safely descend into the wonders of his caverns full of
ancient fish, dinosaurs and prehistoric humanoids before being
washed out in a watery eruption of Stromboli volcano in Italy.

In fact, Verne was not up-to-date, or perhaps took poetic
licence, with Davy's views to frame his story. There was another
theory in play in the nineteenth century about the source of
Earth's inner heat, a theory divorced from chemical reactions
like burning. These ideas also had a deeper lineage, tracing back
to include the French mathematician and philosopher René

Descartes. In his 1644 *Principles of Philosophy* he conceives of 'this Earth which we inhabit' as formerly a star like the Sun only much smaller with the essence of Earth's inner Sun bleeding from its centre. Although, for Descartes volcanoes were not connected with this deeper energy and remained mountains 'notorious for frequent conflagrations', fuelled by sulphur or bitumen, this notion of primordial energy held in the Earth since its celestial birth and thence an alternative source of heat to drive volcanism was an important influence on later thinking. By 1828, more than 30 years before Verne's thrilling tale, Davy himself had decided that evidence from mines and hot springs made it 'probable that the interior of the globe possesses a very high temperature' and conceded that the proposal that the centre of the planet was hot and fluid offered a simpler solution to the phenomenon of volcanic fires than his reactive metals. By the end of the nineteenth century, geology as a field of human endeavour was established within the Western world and it was widely agreed that volcanoes were not driven by burning. Volcanic 'smoke' was realised to be condensing steam or other vapours rather than combustion fumes and the 'flames' were instead related to glowing hot molten material. While solving some mysteries regarding the observations of volcanic eruptions and their products, this moved the problem from locating internal fuel and other reactants to working out the source of the molten rock within the Earth, a question intimately entangled with our planet's inner structure.

In *Principles of Philosophy* Descartes, like Kircher just a couple of decades later, penned an imagined cross section of the Earth. Descartes' image contrasts with Kircher's in both style and content. There is none of the drama of blossoming inner fires or the poetry of ships or billowing winds. It is a clinical depiction of concentric shells labelled, like many science diagrams today, with letters denoting the regions from the outer layers of 'air, stones, clay, sand and mud', and water to the very thick and heavy internal layers from which metals originate and the innermost incandescent material in the centre. The nature of these deepest star-like regions of the Earth concern Descartes' exposition 'very little, because no one has ever reached them alive.' True observation of the inner Earth is elusive, even now, and much of what we know remains based on inference. Hidden in the adventure of *Journey to the Centre of the Earth* is not the explanation for how our world works, but perhaps instead an allegory for the difficulty of actually knowing what is going on beneath our feet.

The Earth is very deep, with 6,400 kilometres separating us on the surface from its centre within its core. A colleague in my department in Oxford had a cartoon pinned outside their office door for a while. It was a stick person standing on a cross section of the Earth with a caption that read, THE THING ABOUT THE CORE IS THAT IT IS ALWAYS RIGHT BENEATH US. Every time I needed to use the printer I would pass it with a twinge of vertigo imagining the planet falling away beneath me in storeys, stretching roughly as far down from me where I stood as the cities of Minneapolis to the west and Lahore to the east as the crow flies. Sometimes when I consider what is going on inside a volcano I try to imagine being there inside its magma storage system, but I inevitably fail. As you go down into the Earth not only does the temperature rise to thousands of degrees Celsius, but with the weight of the planet above, pressures rapidly build to unimaginable levels – thousands to millions of times that on the surface. Even at the relatively shallow levels of a volcano's plumbing, it is just too far outside the pressure-temperature domain of my experience here on the surface. Without being able to go there or truly imagine it, how can we know about this great expanse beneath us? Sadly, there is no runic code to break and no tunnel into Earth's inner workings to be found. Even today our deepest mine shafts are only a few kilometres deep and our deepest boreholes just over 12 kilometres. We barely begin to scratch the surface.

That said, although shallow, incursions beneath the surface by miners did play an important role in understanding our planet's inner nature. By the mid nineteenth century, miners' reports of temperatures increasing with depth had moved beyond anecdote. Measurements had been made at many locations around the planet, recording how steep this increase in temperature was with distance as one went further into the Earth's bowels. Tracing this thermal gradient led to the hypothesis that tens of kilometres into the planet, temperatures would exceed the known melting point of rock and that there would therefore be a molten internal ocean underneath the solid crust. When I think about gazing in on Masaya's roiling magma, I can see how this notion was appealing. Masaya's vent felt like a skylight window onto the secret inner workings of the planet, and there is a tempting simplicity to this explanation for volcanism. But, armed with the edifice of the present-day sum of our human knowledge, I knew that this was not true and that the liquid part of the upper inner Earth on show in the tropical dusk was special rather than pervading.

To make sense of the Earth's deep interior we have to use the clues that we can gather from our rather remote perch here on the surface. We can take what we know about how rocks behave above ground and draw trends to make predictions about how things will be as the temperature and pressure increase deep down. We can test these predictions using experiments or observations. Since the early twentieth century, scientists have developed equipment like diamond anvil cells that allow tiny capsules of rock to be subjected to great pressure and temperature (by squeezing them between the tips of two diamonds and heating them using laser light) to study what happens to them at conditions like those deep within our planet. Sometimes deep rocks are thrust back up onto the surface and give us hints;

messengers from Earth's depths. The deepest exhumed rocks that I have ever seen in situ were in the Dabie Mountains in China. Spilling out of a steamed-up coach onto a riverbank it was hard to appreciate how truly special the rocks were just by looking at them. It is only under a microscope that the tiny diamonds brought up from Earth's depths are revealed – minute trophies of the rock's journey to extraordinary pressures and back again. However, the deepest rocks so far recovered are only from hundreds of kilometres down, so while remarkable they still only plumb a fraction of the planet's depths. Further, although they contain lots of important information, the very fact that they are back at the surface makes them special with somewhat unusual histories rather than typical examples characteristic of the bulk of the solid Earth that remains deep beneath us in perpetuity. This means that valuable as they are, by their very nature such exhumed samples are not equipped to tell us the whole story about what is happening underground.

To understand the vast unseeable mass of the inner Earth we need to use many lines of inquiry. Clues come from large-scale measurements of phenomena like our planet's magnetic field, orbital movements, behaviour in response to tidal forces and earthquake signals. In the 1680s Newton used his new theory of gravity and the celestial dance of the Earth with other planetary bodies like the Sun and Moon, charted since the time of the Babylonians, to calculate its average density. These calculations put bounds on the density required for the bulk of Earth's unseen interior and showed that it must be greater than the crustal rocks at the surface. But this average density alone did not determine whether the interior is mainly solid or molten and other evidence was needed to fill in these vital details about the inner planet. The lack of solid-Earth tides as big as those of the oceans, and the details of how Earth wobbles and spins on its axis as it orbits the Sun, were evidence against a shallow internal shell of molten magma and led to the conclusion that the planet's outer solid layer is far from superficial. One of the masters of these celestial calculations, Lord Kelvin, apparently liked to demonstrate his ideas by spinning a hard-boiled egg and a raw one in front of his audience. The solid egg could spin on its axis while the raw egg would flop around as the fluid flowed within it. According to his calculations in 1863, Earth was a hard-boiled egg, more rigid than glass at least down to 4,000 kilometres.

Experiments showed that increased pressure could cause rocks to remain solid even when they became very hot and should otherwise melt. But perhaps the most definitive information about the inner planet came in the second half of the nineteenth century through the study of the transmission of earthquake waves through the Earth. I remember attending a lecture about earthquakes as a graduate student and being impressed by the fact that after a large earthquake, the whole planet rings like a giant inaudible bell. Remote as the danger of a large earthquake seemed sitting in that lecture theatre in Cambridge, many parts of the planet are unfortunately regularly prone to seismic hazards. Employed as a professor of mining and geology in Tokyo, John Milne and a group of fellow British scientists became the rather unlikely founders of the Seismological Society of Japan after a large tremor struck Yokohama in 1880. To study the way the Earth vibrates after earthquakes this group invented a simple horizontal pendulum seismograph. When an earthquake strikes, vibrations or waves ripple out from its source through and around the Earth. Those waves that travel through the interior of the planet from the earthquake to the observer convey information about the material that they have passed through if it can be correctly decoded.

Although these days great computing power is thrown at interpreting the subtleties of the messages contained within these seismic waves, some of the very fundamental observations are more easily grasped. Seismographs like Milne's were able to see two types of waves travelling through the Earth. First, they would see the wiggles of primary (P) waves: formed of patterns of compression and expansion travelling through Earth's fabric in a similar way to the pulsing of air that we hear as sound. Later to arrive would be the slower, snake-like secondary (S) waves. These are like ripples on the surface of a still pond or a Mexican wave going around a sports stadium. A crucial difference between these two types of wave for the study of the Earth is that while P waves can travel through solids and liquids, S waves can only travel through solids. For an S wave or a Mexican wave to travel through a substance or around a stadium, no individual atom or person is moving in the direction of the wave. Instead, in the case of the sports fans, they are just standing up and sitting down while the wave itself moves from person to person circling the stands. To achieve this sort of wave you must have something that pulls the displaced atoms or people back down after they have moved out of position. In a Mexican wave this is human volition and muscles. On the atomic scale it means that these S waves can only travel through an ordered solid, where atoms, pushed out of their position in the solid's gridded seating plan by the wave's energy, will be yanked back into line as if on a bungee cord. In a disordered liquid or gas where atoms or molecules are milling around like a chaotic crowd there is no structure to pull the next-door molecules up or back in line as the energy passes through. The very fact that S waves made it through the inner Earth told scientists that it was solid.

By 1906 seismologists had used their seismometers to trace wave pathways from distant earthquakes deeper still through the Earth and found that the traces that passed below a certain depth lost their S waves. This showed that the Earth was indeed molten inside but only below 3,000 kilometres. These seismic traces drew a picture of the inner Earth defined by concentric shells: a 5–70-kilometre-thick crust, a mantle extending to about 3,000 kilometres' depth and deeper still the molten outer core hosting, as was later discovered by the Danish seismologist Inge Lehmann, a solid innermost sphere.* The idea of volcanoes directly tapping into some internal magma ocean was ruled out once and for all.

This left a conundrum: how could liquid magma be generated from the solid rock of the planet's upper reaches? The most obvious answer came from the same experiments that had shown that hot rock could be solid at the great temperatures and high pressures within the Earth: if increasing pressure could raise the melting point of rock by squeezing its atoms together, then releasing that pressure and letting the hot, buzzing atoms spring apart would cause it to melt. Most of us live much of our lives within a relatively narrow range of altitude and therefore pressure. This means that we do not often need to trouble ourselves too much about the effect that pressure has on the behaviour of substances around us. But sometimes we leave our normal pressure envelope, probably most commonly with a trip up a mountain.

During my PhD I spent two weeks in the small village of Talabre in the Chilean Atacama Desert, a place to which we will return in Chapter 7. At an altitude of roughly 3,500 metres, the atmospheric pressure is only about 65 per cent of that at sea level. You feel it physically and to begin with even small tasks like carrying equipment from the car leave you panting. I had expected this. I had not expected the effects on our culinary endeavours. At these lower atmospheric pressures, the water molecules can more easily overcome the forces holding them together in their jostling liquid form to escape as free gas molecules. This means that the boiling point of water is about 12°C lower than by the sea. Despite being a group of scientists, our inexperience at managing this particular physics meant that we

---

* Lehmann's discovery was presented in her famous 1936 paper simply entitled *P'*, in which she interpreted certain unexpected P wave arrivals as reflections from a solid inner core.

ate rice as slow-cooked mush for the duration of our stay.

You can demonstrate this in a lab by bringing a beaker of room-temperature water to the boil by sealing it in a vessel and pumping out the air. In the same way that a drop in pressure can cause a pan of water to instantaneously boil without changing its temperature, moving a batch of hot rocky mantle to a shallower depth within the Earth has the potential to generate molten magma. By the early twentieth century, experiments and theory pointed to changes in pressure as a plausible mechanism for creating magma from solid rock. But initially geologists were not able to make sense of the scales of decompression needed to drive Earth's magmatism. What might cause great swathes of the solid mantle to rise up within the Earth to lower pressures and subsequently melt? As it was to turn out solving this conundrum is tied up with important issues surrounding heat flow and loss through and from the planet. This problem became all the more pertinent in the first half of the twentieth century with the discovery of a powerful new source of energy – radioactivity – permeating Earth's inner fabric, and held within the tiny nuclei of the very atoms making up its vast rocky bulk.

In my grandparents' house there was a strange picture on the wall. I liked to pause to look at it when I was a child. It was of a man and a woman, a couple, the woman's hand resting on her partner's shoulder. They were surrounded by weird glassware and jars but what drew my attention most was the vial of substance held aloft by the man from which crepuscular rays of light emanated. Underneath was a single word: 'Radium'. As a child I found it mysterious and intriguing but was not sure who they were or what the image meant. For some reason I uncharacteristically never thought to trouble anyone and ask about it. I forgot about it for many years but then as an adult, with the benefit of all knowledge only a few clicks away, I came across it again by chance and found out that it was a faded print of an illustration of Marie and Pierre Curie's discovery of the highly radioactive element, published in *Vanity Fair* in December 1904. I do not know where the picture is now and I never did ask my grandparents where they got it from.

The Curies were at the vanguard in researching and understanding the phenomenon of radioactivity. These new observations rocked the foundations of physics at the time and also had profound implications for our understanding of the planet. Radium is about a million times more radioactive than uranium, and so energetic in its decay that the heat emitted

Radium: Marie and Pierre Curie, from Vanity Fair, *December 1904*

can be measured with a thermometer. The physicists Ernest Rutherford and Frederick Soddy showed that a lump of radium could melt its own weight in ice in an hour. Measurements of varieties of volcanic rocks from across the globe indicated that radioactive elements like uranium (which decays into radium, among other elements) and thorium were common. Estimates of their total abundance in the Earth suggested that this newly discovered process of decay accounted for a significant proportion of Earth's heat and present-day estimates reckon that this radiogenic heat and the primordial energy still lingering from its formation each account for roughly half of the energy flowing from the planet's interior to its surface. The discovery of radioactivity challenged the then current understanding of the world in many different ways, from the minute, namely its subatomic structure, to the global, and the source of heat in the very rocks beneath our feet. This new theory shifted our ideas about Earth's heat again, and now Descartes' inner cooling star was augmented by a slow-paced nuclear reactor, with the random disintegrations of its trillions of radioactive atomic nuclei bleeding energy through the rock over eons.

All this extra inner heat fed into another debate about Earth's behaviour, however, namely how all this thermal energy was getting from the great depths of the planet to be lost through the surface. Within a totally rigid solid, heat can only travel by conduction – and rocks are not efficient conductors for heat energy. If conduction were the only heat loss mechanism in play the Earth would not have lost significant heat from the full depth of its mantle even over the course of its entire lifespan. But if the mantle could somehow move and mix by a creeping rise and fall of convection, a different scale of heat loss could be made to work. While conduction involves the passing of energy from one molecule to the next, convection results in hotter, more buoyant material rising and taking its heat energy with it. Sometimes I will watch convection play out in my breakfast coffee when I add the milk. Chalky blossoms of cooler milk swirl in eddies, sink and then rise to the surface, to fall back again. This miniature convective storm within a coffee cup, while a moment of morning meditation also, it turns out, holds parallels with the complexity of our planet's interior ever present beneath my feet.

Movement of the inner Earth on this scale explained other things too. In 1912, the German meteorologist Alfred Wegener proposed that the Earth's great crustal plates moved over time, based, among other lines of evidence, upon observing the jigsaw fit of Africa and South America's coastlines and rocks. In the 1920s, scientists like Arthur Holmes suggested that mantle convection might offer a combined explanation of Earth's heat loss and Wegener's theory. The idea of a convecting solid mantle was still hard to stomach for many geologists at that time, but in the early 1960s, observations like the great lines formed by the locations of volcanoes and earthquakes on both continents and the sea floor and, perhaps most persuasively, the symmetrical zebra stripes of rock with different magnetic properties on either side of ocean ridges helped to provide us with a unified theory of plate tectonics.

Our current best theory imagines the solid mantle creeping slowly beneath us in vast circulations, rising and sinking, and keeping rough pace, as it happens, with the growth of human fingernails. Continents dance over geological time: coming together and splitting apart. If we run time backwards, the world map of past geological eras rapidly becomes unrecognisable to human history. If we could fast-forward the clock we would see that our geographical truths are not immutable, that in time the shapes of our oceans, continents and countries will look very different as we cast our gaze far beyond ourselves into Earth's long geological future.

Among the many achievements of the 'plate-tectonic revolution' was a way to link simple observations of the locations of global volcanoes to the underlying processes leading the mantle to melt and magmas to form. From antiquity, it had been noted that volcanoes often occurred as islands, or at least not far from the coast. One of the earliest Romans to discuss volcanoes, the poet Lucretius, attributes Etna's activity directly to the sea. In his account, air, water, sand and stones were rolled into caves within the hollow mountain by the waves. Although the link between volcanoes and the sea is, of course, not quite as poetic or direct as this, as our maps stretched out beyond the classical world it became clear that most of the world's land-based volcanoes traced the margins of the continents or formed chains of islands striking out from the land.

Harder to see, but no less impressive, are the great chains of volcanic mountains and vents deep under the oceans. These ocean ridges sometimes peep out above the sea, forming volcanic islands like Iceland or the Azores. But mainly their existence remained hidden until oceanographers began to survey the ocean floors in the nineteenth century. These great lines of submarine and land-based volcanoes mark key moments in the life cycle of our tectonic plates. They show where ocean floor is born and where it dies. This slow-scale drama, driven by the engine of Earth's internal heat, plays out across the planet, but the Pacific is a particularly good place to watch it unfold and it allows us to draw back to Nicaragua and to pause again atop Masaya to take in the view.

From Masaya's summit you cannot actually see the Pacific Ocean, which lies 40 kilometres to the south-west as the crow flies. You can, however, see the neighbouring volcanoes in both directions: the mighty cone of Momotombo to the north-west and the broader shape of Mombacho to the south-east. If you could take a giant step from Masaya to Momotombo you would then gaze on towards the young, cinder cone of Cerro Negro (Black Hill) and beyond it the taller fuming majesty of Telica. From Mombacho, the view to the south-east over Lake

Nicaragua is punctuated by the cone of Concepción volcano on Ometepe Island. From Masaya, volcanoes reach in both directions like links in a volcanic daisy chain along the fine limb of Central America. These Central American magmatic mountains take their place in a broken chain that circles the Pacific. This 'ring of fire' stretches from the tip of Patagonia through the Americas, the arc of the Aleutian Islands ploughing through the Bering Sea into the Russian peninsula of Kamchatka, Japan, the Philippines, Indonesia, Melanesia and finally New Zealand. This great fence of volcanoes corralling the Pacific is perhaps one of the most impressive expressions of volcanism readily visible to us above the waves, but the processes that formed it started far away from the continents, at the hidden ocean ridges.

Deep beneath the Pacific waves, new oceanic crust is born. The giant sutures of the East Pacific Rise and the Pacific-Antarctic Ridge bow all the way from the Gulf of California towards Victoria Land in Antarctica. In contrast, the shorter Juan de Fuca Ridge is a broken mirror of a short sweep of the coastline near the US-Canadian border. As the Earth's tectonic plates pull apart, in these places, they unload the hot rock below and it wells up to lower pressures. Going back to my weeks eating mushy rice in the high Atacama, it is as if one was carrying a pan of very hot (but not boiling) water up a mountainside to watch it spontaneously boil. In this case, as the pressure drops the mantle melts and the buoyant magmas find their way upwards, cooling at shallow levels to add new ocean crust to the spreading plates on either side, or billowing onto the ocean floor, cooling rapidly on contact with seawater and freezing as pillow lavas to form new sea floor, or building larger seamounts or volcanic islands.

Importantly, these processes also draw seawater into the new crust through cracks as the plates pull apart and heat it up to as many as hundreds of degrees Celsius. Some of this water will get pumped back out through chimney-like structures as

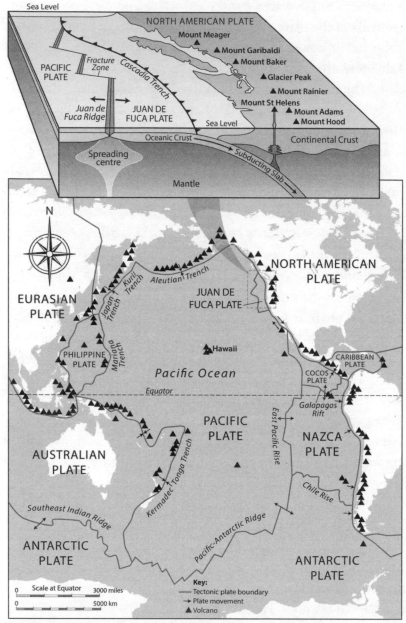

The main map shows the Pacific tectonic plate with some of the key features marked.
The inset above it shows a cross section through the Earth's crust and upper mantle showing the tectonic processes in play associated with the small Juan de Fuca plate as indicated by the dashed rectangle.

plumes of mineral-rich water, called black or white smokers. But some of these water molecules will get bound into the hot new crust of the plate as it forms. The great internal heat engine of the Earth powers these oceanic plates away from the place of their creation, filling the space with new lavas and new crust behind them, like a gigantic conveyor belt. As it travels, detritus from above – for example, layers of shells, sand, mud or even volcanic ash – settles onto the ocean's rocky bottom, binding more seawater with them as they form, layer upon layer of murky sediments along for the ride. This layer cake of sediment and frozen magmas underlies the seabed deep beneath the rolling surf or raging waves of our blue planet. But it is not forever. Very little of the sea floor is older than 200 million years. As sea floor is born at the ocean ridges where plates part, it dies where plates collide. The volcanoes of the Pacific ring of fire rise like monuments to this destruction.

Where oceanic plates push into another plate, especially one topped with lighter continental crust, gravity can pull the denser ocean floor down subducting it back into the Earth. The great trenches in the ocean floor all around the Pacific mark where this process of subduction is occurring. Many of these are among the very deepest parts of the world's oceans, going further below the waves than Mount Everest rises above them. Strange life is found in these dark, high-pressure waters, like mysterious shrimp-like scavengers and bizarre translucent sea cucumbers. It is not just the volcanic rock of the oceanic crust that sinks into the Earth's depths, but also seawater – some that was bound to the crust in the heat of its birth, and some in the plate's mantle of sediments. As the slab grinds downwards, the rising temperature and pressure inside the planet start to stew the old sea floor. During this cooking process at least some of the water taken down with the sinking plate is driven off. It percolates up into the mantle above and changes it. The effect is a little like pouring salt on a frozen pavement. The salt atoms disrupt the freezing process, pulling down ice's melting point and turning lethal sheet ice into mush. Similarly, adding water to the mantle turns it mushy without needing a change in temperature, generating magmas, some of which will go on to reach the surface and build volcanoes.

As I gaze down into the magma in Masaya's crater, I'm watching the culmination of a process and a journey. We cannot ever directly see it, but 100 or so kilometres beneath volcanoes like Masaya, beads of molten mantle squeeze together to form buoyant balloons and channels of magma that start their journey towards the surface. Not all of this magma will make it. Some will freeze again, helping to build up Nicaragua's crust from beneath. All of it will be changed by the journey. The portion of it that makes it up to meet us at Masaya can, as we have seen, entrance not only volcanologists like me but also the tourists who come to gaze over the rim as it illuminates the night, peering in at the roiling remnants of the great tectonic processes playing out across the expanse of the Pacific.

In quiet moments between measurements in the gassy car park, the glow coming from Masaya inspires me to imagine the great depths of Earth below me, and the huge scale of the process of volcanism. It is a view that has meant many different things to the many different people who have peered into the volcano's mouth, from the first indigenous peoples to observers like me in the present. It is a view that has been different but the same over the centuries, both in the scene and spectacle that the volcano presents, and the way we understand it. I stand on the shoulders of generations of thinkers and scientists before me who asked 'why?' when then they gazed at the molten Earth and have collectively come to our current theories of Earth structure and plate tectonics. As the Sun sets, I might stare along its path towards the invisible Pacific Ocean and imagine the sea floor creeping imperceptibly towards me before sinking into the depths. It is a struggle to comprehend the fathoms of crust and mantle beneath my feet and to take one's mind to where the glowing lava was born or the long, slow processes that brought it to our gaze. It is a reminder that although we have come a long way in understanding the origins of volcanism, many mysteries remain.

One of our missions on that first trip to Masaya was to measure how much gas was coming out of the volcano. We were doing this with a then relatively new instrument, a cheap, lightweight spectrometer about the size of a pack of cards. Inside this box of tricks, an optical grating splits light into its separate wavelengths – a spectrum – like a prism spreading sunlight into colours to make a rainbow. Clever electronics then turn this spectrum into a digital record of intensities. By comparing subtle differences between the spectrum of light travelling through the volcano's gas plume and that coming through the clean air, an algorithm can count the molecules of sulphur dioxide gas emerging from the Earth.

Things had not always gone smoothly with this new-fangled instrument. It had been in one of the bags lost in transit, and after we had eventually reclaimed it from Nicaraguan customs, we found that the spectrometer was partly broken. A hasty superglue fix was improvised so that the telescope would focus enough light to take measurements. High tech as science might appear from the outside, in volcanology at least, things are often reliant on glue and duct tape to keep the show on the road. But once we made it work, its small size and light weight meant that we were able to tape the spectrometer to a helmet so that we could walk under the plume rather than needing a car or other vehicle, a convenient way to take readings even if it meant looking something like a cyborg.

Our measurements confirmed what others had suggested. At Masaya, as with many other volcanoes, there is much more gas coming out than can be explained by the volume of lava that it erupts over time. As we will see in later chapters, volcanic gases are dissolved in magma and fizz out as it reaches shallower depths. This mismatch between the rates of gas emission and lava outpourings at Masaya tells us that most of the magma driving the activity must sink back into the depths once it has exhaled its buoyancy-giving gases. Right from my first glimpse of Masaya's crater glow at sunset, I was aware that we were seeing only a very small part of a vastly complex picture. It was not just a matter of wondering where the molten Earth had come from, but also what the consequences were when it drained back into the bowels, out of sight – likely sinking deep to find cracks or weaker areas to spread out and gradually cool, building the crust from below and heating the rocks around it. In our quest to comprehend volcanic behaviour, understanding how and why the Earth melts is just the first step, and of course, when the magma breaks out rather than retreating below, the consequences can be far more destructive.

Celebrating my twenty-fifth birthday on a volcano was certainly different. I remember that, as a treat, I was allowed the last can of tinned fruit from the cupboard for breakfast, that I spent a hot, dusty day falling in drainage gullies behind the volcano while putting sampling equipment out, and that some friends from Cambridge had smuggled a present into my luggage, which turned out to be a survival handbook including a section on how to endure a volcanic eruption – a gift that likely felt far more edgy when opened within sight of the volcano than intended.

Fortunately, Masaya has, for the most part, been a relatively benign volcano in recent years. But, as the thick layers of ancient ash in some places around the volcano tell, this has not always been the case, and there is always the chance that the gas wafting gently out of its vents could one day again result in a massive and destructive explosive eruption, a scenario that volcanologists would, of course, aim to anticipate. The story of how magma behaves after its deep genesis is a tangled web of molecular

Masaya volcano and the plume emanating from its active Santiago crater viewed from the north-west in 2017. The steep cliff to the right of the photo is the volcano's old caldera rim.

networks, crystals and bubbles. Activity at different volcanoes can have very different characters: from relatively approachable lava lakes in places like Hawaii to the deadly pyroclastic surges from Vesuvius that engulfed Pompeii, and much more besides. Eruptions from just one volcano can come in a wide range of styles, even over relatively short periods of time: from magma stalling and stowing in the crust, oozing or squeezing out as flows or domes, to sudden explosions blowing plumes up into the atmosphere and raining hot material down around it. These very different volcanic outcomes between volcanoes and between eruptions are governed by factors both locked in at a magma's birth and imparted during the different pathways they find on their way towards the surface. Exploring more about these processes and their consequences takes us onward on our adventures into Volcanoland. On our very human scale, the legacy of molecular-scale processes in the Earth's oceans, mantle and crust can mean the difference between life and death, deliverance and disaster. But as with our imaginings of the deep inner Earth, there is much that we have learnt that can help us understand why volcanoes behave as they do, even if our predictions of these complex systems remain far from perfect.

Chapter 2

Journey to the Surface of the Earth

Why do volcanoes erupt in different ways?

A lthough ominous, the sloshing lava and crater glow at Masaya does not speak of volcanic cataclysm and catastrophe. Here the molten Earth is something to peer into and wonder at and the risk, though present, is low. But human history tells a very different tale of the dangers of these portals between the inner Earth and its surface. Volcanoes are well known as agents of destruction. Powerful stories from antiquity, like those that I encountered during my first childhood visit to the Bay of Naples, leave a deep impression on our collective psyche for good reasons. Since becoming a volcanologist, I have studied the rocky products still preserved in the landscape from several very large-scale explosive eruptions and I will describe how we reconstruct such events in the next chapter. Usually the culprit volcano has been in a state of relative repose when I have visited, seeping gas perhaps, imperceivably warping or quivering with unfelt seismicity, but without raw and blatant manifestations of their explosive potential unless you take pause to interpret the messages in the ancient rocks. But in 2009, I visited Santa María volcano in Guatemala, the site in 1902 of one of the largest eruptions of the last 200 years. Not only is the land here still scarred by this massive eruption, but like Masaya, Santa María is continuously active today. Although its current behaviour is fortunately nothing like the great 1902 tragedy, its frequent small explosions from its latest vent complex (known as Santiaguito), that send wispy plumes of ash a few kilometres into the sky and carve out a 'no go' zone around its active cone, are in striking contrast to the more contained roiling machinations that we gazed into within Masaya's vent.

A small and short-lived explosion from the Caliente vent of Santiaguito in 2009. Santa Maria volcano is in the right of the image with the 1902 eruption scar still visible.

Masaya and Santa María sit in the same subduction zone, about 650 kilometres apart, along the line of volcanoes that traces the sinking of the Cocos tectonic plate beneath the Caribbean plate. The magmas driving these volcanoes therefore share similar origins and yet while Masaya's historic activity has been punctuated with the liquid 'fire' of lava lakes and flows, Santa María's has been far more explosive and threatening. How can we explain such different styles of volcanic activity despite such ultimately similar origins? And more appositely for the populations living nearby, how do we understand why some volcanoes have greater explosive potential than others? The answers to these questions lie in the varying ways that magma is born and then moves from its deep source to the surface. Insights from many thinkers over the centuries have given us a greater understanding of these processes. Understanding the very molecules within the rocky melts – their different chemistries – is critical to predicting how a magma will tend to behave once its magma reaches the surface. Contemplating the events of 1902 and visiting both Santa María and Masaya in the present day offer an opportunity to explore how this science of molecular-scale magmatic change can mete out such different consequences for the people who live with these volcanoes.

The year 1902 was a busy and destructive one for tectonic forces in the West Indies and Central America. In April a magnitude 7.5 earthquake rocked Guatemala and the surrounding region, with the destruction centred around the Guatemalan city of Quetzaltenango (or Xela in Mayan). Records are sketchy, but hundreds were killed, and nearly all the churches in western Guatemala were heavily damaged or destroyed. Hot on the heels of this tectonic tragedy came another on 7 May, with the eruption of La Soufrière volcano on the Caribbean island of St Vincent killing at least 1,600 of the local population. Unthinkable as it might seem, this catastrophe was horrifically

eclipsed the very next day when Mount Pelée erupted on nearby
Martinique. The deadly collapse of a super-hot flow of volcanic
gas and ash down the mountain's slopes killed almost 30,000
people in the space of a few minutes and reduced the city of

*An image of one of the ash clouds over Martinique characteristic of the
smaller-scale activity that followed the terrible eruption of Pelée in 1902.
From* A. Lacroix La Montagne Pelée et ses Eruptions *(1904).*

Saint-Pierre to ruins – the worst volcanic disaster of the twenti-eth century, as it would turn out.

But the year's drama was not done, and on 23 and 24 October Santa María volcano in Guatemala blasted its side off in a mas-sive eruption killing thousands. Ash was reported as far away as San Francisco, a distance of almost 4,000 kilometres. Oblivious to the turn of the human calendar, the Earth had nonetheless seen the new century in by roaring and raging.

Guatemala was still considered relatively remote and inacces-sible by many Westerners in 1902. The country was reachable only by steamer from San Francisco or Panama – a circuitous and expensive route, especially for those coming from Europe. However, the Swedish American Gustav Eisen, a man with the rather lengthy job title of 'Curator of Archaeology, Ethnology, and Lower Animals at the California Academy of Sciences', happened to be there to witness the events of the eruption. His account is rich in information and reflection although he is, of course, an outsider and does not write from the perspective of those whose homes and communities were devastated. Even before the eruption, he was already entranced by the country's volcanoes. The view of Guatemala's imposing and spiky volcanic backbone, set back 60 kilometres inland from the Pacific, mir-roring the coast, had impressed him from the steamer on his voyage there. He recalls them as 'the most characteristic, as well as the most interesting and beautiful features of Guatemala'. Geography certainly does seem to contrive to show off these peaks from this ocean vantage. The volcanoes perch on the southern border of the country's highlands so that they rise from sea level to their full majesty – two or three kilometres in altitude – above the coastal plains and coffee plantation-covered foothills. Their presence has impressed humankind as far back as our eyes have gazed upon them. According to the ancient Mayan creation myth preserved in *Popol Vuh*, Santa María (named as Xcanul in the K'iche' language), along with other

Guatemalan volcanoes like Fuego and Agua, was among the great mountains called into existence in a single night.

Arriving into Guatemala City's busy airport today certainly feels less romantic than steaming along the coast by ship. Nonetheless I can vouch for the stunning views of the volcanoes on the drive from the Guatemalan capital to the coastal plain highway. When I visited in 2009, I had the privilege of being picked up for the drive by Gustavo Chigna, one of the small but dedicated team of local volcanologists charged with monitoring the country's volcanoes. It was January and I felt an immense joy to be out of the UK winter and bathed in the colours and light of the tropics. From Guatemala City the road rose up, traversing a fault scarp, towards the former capital Antigua. We chatted about the great stripes of volcanic rocks exposed in the road cuts, and paused in Antigua just long enough for me to prove once again in the local artisan market that I really am no good at bartering.

On towards the coast from Antigua, the road plumbs the valley between two imposing volcanic cones, Fuego (Fire) to the west and Agua (Water) to the east. Fuego has been very active in recent years – tragically so in 2018 when pyroclastic flows buried local towns, killing hundreds, perhaps thousands of people. In 2009, its activity was more benign and it seemed to salute us with the ashy puff of a small explosion as we caught our first glimpse from the road. I felt more than a twinge of guilt for having smiled in delight at Fuego's beauty then as I watched the terrible scenes unfold from the safety of Oxford almost a decade later. Eisen was certainly right about the beauty of the volcanoes, but it is a dark beauty and there is always an edge, a catch. When, in awe, they take your breath away, there is always the risk that one day they won't give it back.

When the Santa María eruption began in 1902, Eisen was in the small town of Rabinal. Despite being more than 100 kilometres from the volcano, he was awakened in the night by what he thought to be 'heavy cannonading or firing of bombs'. He initially put it down to celebrations of the Fiesta de Minerva, but by morning he was convinced that what he was hearing came from one of the country's volcanoes and telegrammed Guatemala City to find out which one. Later he scaled the highest nearby hill in the hope of seeing something. As he climbed, the noise from the explosions grew more and more intense. By sunset the underground detonations trembled through the hillside such that it felt 'ready to burst', and at times he had to hold on to the rocky terrain in order not to be thrown over.

I have never been near an eruption even approaching the size of that in Guatemala in 1902, but the way that Eisen describes the throbbing Earth resounds with my own experiences of small explosions. Although unrelated to volcanoes, a Plato quote captures the sensation for me: 'Sound is a blow delivered by air, through the ear, on the brain and the blood, and transmitted to the soul . . . ' These days we know more of the process: a pressure wave travelling through the air resonates our ear drum, triggering electrical nerve impulses to the brain. This idea of 'a blow delivered by air', though, expresses the way this loud energy can reverberate right through your body, sometimes to the point of feeling violent. Volcanic explosions take this feeling to a more profound level. The intense blow is not only delivered by the air but comes at you through the ground, from deep within the Earth itself. It feels as if the sound and sensation is pulsing into your being from all directions. I am not surprised that Eisen needed to brace himself against the hillside.

In 1902, the Guatemalan government was not keen for news of the eruption to get out. Eisen received a negative response to his first telegram and later received dispatches suggesting that a volcano in Mexico was to blame: a claim that no one around him believed. After a tricky journey severely impeded by washed-out mountain roads, Eisen eventually arrived in the disaster zone around Santa María. Local accounts described days of darkness, echoing underground explosions, lightning, and sand and mud raining from the sky. Steamer captains out at sea had used their sextants to estimate the height of the great ashy column as punching about 30 kilometres into the heavens.

The electrical phenomena during the eruption particularly seems to have made an impression on Eisen. For 24 kilometres from the crater, trees had been destroyed by lightning, and in the farms – *fincas* – on the slopes of the volcano, electric sparks crackled between people's feet and the ground as they picked their way over the ash-covered verandas. Others described countless curved lines of red and green electric discharges. Strange as they may seem, these charged phenomena are quite common during violent volcanic outbursts. When rocks break and collide in the high energy of a volcanic explosion, electrons are shed from or swapped between the fresh surfaces, loading the ash and gas with electrical charge. High up in the cloud, volcanic water can freeze and become volcanic ice and the whole mucky mixture can convect and charge like a dirty thunderstorm. These mechanisms build up electric imbalances, air's resistance to electrical current flow is broken down, sparks fly, lightning arcs and sometimes strikes.

It was two weeks before the clouds cleared sufficiently for Eisen to view the remains of Santa María. The main cone of this beautiful volcano, previously a near perfect pyramid, now had a deep elliptical scar facing the ocean, rent with steaming vents and cracks – or fumaroles to give them their more technical name. A great bite had been taken out of the mountain. As he looked upon this scene for the first time, he saw vast thunderous quantities of rock constantly falling into this new crater and Eisen thought it probable that the very top cone of Santa María would likewise topple in. The surrounding landscape was cloaked in a thick mantle of white and grey 'sand' as far as the eye could see. 'Of the tropical and impenetrable vegetation which once clothed every inch of ground which had not been previously cleared for plantations nothing was left except bare trunks and branches.' The landscape was transformed. Close to the volcano, farmhouses were completely buried. Some ravines and hollows had been filled in and new hills built from the debris elsewhere. Eisen likened the scene to a snowy wasteland. The volcano had bleached the wet, warm tropical greens from the landscape's palette, scrubbing everything into an anaemic barrenness. The ash fall could be traced hundreds of miles away with chunks of pumice caught in the branches of trees and rafting in the ocean, lapping on the waves and hiding the water from view.

October is the end of Guatemala's rainy season and water was already at work on this new landscape. Billions of raindrops hammered down from the heavens, cementing the ash in places and etching out fine feather-like networks of coalescing water channels in others. Coursing their way through the friable material, these trickling barbs of water rapidly came together into torrents to carve out great new trenches through the ashy landscape. Large rivers like the Samalá completely changed parts of their course and almost every afternoon the torrential rains rushing down from the hills swelled their waters in a great surge of 'thick gruel-like mud' caused by the rain's erosive dialogue with the fallen ash and pumice. These terrifying muddy deluges – known to volcanologists since the 1920s as 'lahars' (an Indonesian word for volcanic mudflows) – roared with the noise of distant thunder and swept boulders, trees and dead cattle in their maelstrom.

Pumice, mud, stones, sticks and ash turned out to be effective dam material when wedged in the drainage, forming hundreds of flood lakes. Anyone who has ever dammed a beach stream with stones and sand will know the force of the torrent released when a trickle of water breaches a sandy wall and erodes a path, unleashing the rest to follow in a great rush – 'a great whoosh' as my grandfather would call it when I was a child. In Guatemala in 1902, these same processes on a brutal scale swept away most of the bridges during the first few days of the eruption. Solutions had to be improvised with cages suspended from wires or hammocks replacing some of the crossings so that in Eisen's words: 'the not too timid traveller could pass'. Over a century later, having traversed these ravines on still rickety (though likely far more solid) wire and plank bridges, I can confidently call Eisen's assessment an understatement.

The impacts on the local population and wildlife were profound. Half the Pacific coast coffee crop was lost. Cattle died in the eruption or from having eaten leaves and grass covered in ash or drinking water mixed with mud. Eisen reckoned 1,500 people were killed from falling roofs or being engulfed in debris, although other estimates put the death toll far higher at more like 6,000. On the back of the earlier earthquake that year there were few funds available from the government to help with the recovery, and the economic consequences on top of the human tragedy were crippling, especially for the indigenous population.

The local mammal and bird populations had been decimated. Eisen reports the stench of dead animals buried under the ash accompanying his every step through the tropical forest. While it was a month after the eruption before the first birds – blackbirds – returned, insect life seems to have bounced back more swiftly. Butterflies, beetles, flies and mosquitoes were all noticeable by their abundance despite the blanketing ash. Eisen observed that the zompopos – large Guatemalan winged leaf-cutter ants and a local delicacy for some – had already begun to dig new tunnels. I love the image that he paints of them carrying up the underlying brown soil, now buried a metre and a half down, and placing it on the pale canvas of the devastated surface. I imagine the blossoming brown of their spoil heaps spreading like ink on the white blotting paper of the pale ash, a chromatic rejuvenation, a first step in life reclaiming the bleached desolation.

Unlike eruptions further back in history, we don't have to merely rely on the images conjured by words to witness the events of 1902. Thanks to the then burgeoning techniques of field photography we have a direct pictorial record, too, albeit to my knowledge largely from the perspective of external visitors to the scene of devastation. The wonderfully named Tempest Anderson, a British eye surgeon and keen amateur photographer and volcanologist from York, had been dispatched by a Royal Society Commission to investigate the Caribbean eruptions of La Soufrière and Pelée earlier that same year. With the news from Guatemala he extended his trip to document this eruption, too, collecting testimonies, as well as surveying and photographing its aftermath. His greyscale photos capture a newly swept-out gorge in the River Nima downstream from Santa María, his expedition party resting on the hummocked ridges of the new ash surface and, most impressively of all, the blown-out side of the mountain: the gaping wound of the new

A ridge in the new ash on Santa Maria, taken c. 1906 by Tempest Anderson and published in The Geographical Journal *in 1908*

crater still steaming, the sliced-through stripes of its inner strata laid open, with the bald feather-channelled badlands of eroded ash banks and forlorn stumps of cleared trees in the foreground.

This scar in the once-symmetrical cone looks little different today. The top cone of Santa María remains on its crest, contrary to Eisen's fears. I once spent a train trip through the gentle English countryside intensely flicking between Anderson's crater photo and my own similar view from 2009 in rather last-minute preparations for a talk. Between the two images, divided by more than a century in time, I could trace the summit's constant outline and follow the tumble of the same shadowy ridges buttressing into the crater's depths from the mountain heights. But while the 1902 cataclysm was there in the backdrop in the image from 2009, the action has shifted and the dissected pyramid of Santa María is no longer the main player in the piece. Within the hollow of the old mountain, a new volcano – the diminutively named Santiaguito – has been growing since the 1920s.

There is little family resemblance between the youthful Santiaguito and the older pointed majesty of Santa María's remaining outline. Santiaguito presents a far more complex picture than the steepled symmetry of Santa María that would have mingled with the other peaks greeting Eisen's arrival by steamer along the coast before the eruption. Instead, a century of building has pushed up a row of four dusty grey rocky mounds – lava domes – forming a ridge of interconnected volcanic spoil heaps. Each rugged cupola has built itself up in slow rocky spines pushing up and out, and sluggish blocky lavas tumbling down the rubble-heap sides. Each has a name: Caliente (hot) nuzzles in closest to Santa María's great cleft, then to the west, in sequence, La Mitad (the middle), El Monje (the monk) and El Brujo (the sorcerer).

They are not all active at once and since the 1970s Caliente has been the centre of the action. Viewing the scene from the

Santiaguito volcano observatory in 2009, the smooth cone shape formed by Caliente's rubbly apron of loose rocky debris – talus – was buttressed by the blocky knolls of old lava flows. One flow, from the early 2000s, stretched like gnarled knuckles towards me down the volcano's flanks, before getting lost in the verdant greens of the jungle. There were no active flows when we were there, but several times an hour an explosion would boom out of Santiaguito's summit. Although small by volcanic standards, these explosions sent plumes of pale grey ash a few kilometres up above the volcano and cascades of hot, dusty white material down Caliente's sides. Sometimes stretched into wisps by the breeze, these bright plumes could look almost whimsical: white question marks written by the Earth and wind against a blue sky.

Spiny, rubbly lava domes like these, nestled together in Santa Maria's shadow, are not especially rare across the world's volcanoes. It is not unusual for lava-dome growth to precede or follow large explosive events either. After the fury of its eruption in 1902 was spent, Pelée pushed out a great spine of lava, eventually growing to tower over 300 metres above its crater, like a monument to its victims. La Soufrière volcano in St Vincent continues to interleave such rocky extrusions with more explosive activity to this day. For example, in December 2020 it began to push out a new lava dome in its summit crater. This new dome was over 100 metres high by April 2021 when the volcano shifted into an explosive phase, shooting ash plumes about 16 kilometres into the atmosphere and triggering the evacuation of about 16,000 people. After the famous 1980 eruption of Mount Saint Helens in Washington State, USA, that blasted the side off the mountain and spread ash as far east as the Great Plains, lobes and spines of magma squeezed out to form lava domes within the amphitheatre scar left by the eruption, sometimes resembling arching whale-backs.

Sometimes these lava domes glow red hot at night, especially where rock falls expose fresh surfaces. But the dome-like shape of these structures and the way they grow, often in laboured extrusion of spines or blocks that collapse and crumble as they cool, tells of stickiness rather than fluidity in their movement. Despite being formed from the same basic chemical ingredients, the sluggish movement of this magma can feel hard to materially relate to the writhing, roiling lava lake of Masaya or the freely oozing lava flows in places like Etna or Hawaii. How can we account for such marked differences in the stickiness of these different types of rocky fluids? How can we explain why they flow so differently?

How a magma flows dictates not only whether it will form a runny lava lake or flow versus a sticky lava dome, but also how violently it might erupt. Stickier magmas are also more likely to trap gas bubbles inside them, leading to great explosive eruptions like those that played out across the West Indies and Central America in 1902. In fact, observations of explosive outbursts from the growing spine of Pelée, following the cataclysmic and tragic eruption that so devastated Saint-Pierre, led to important insights into the processes causing volcanic explosions. Studying these dusty blasts led the French geologist Alfred Lacroix, dispatched to Martinique by the French Academy of Sciences, to realise that gas was dissolved in magmas and that explosions were triggered much like the opening of a champagne bottle. Fundamentally a volcano's fate and that of those who live around it is written in its molecular-scale rock chemistry. Chemical bonds dictate how a magma will flow and whether gas bubbles can escape freely or will remain trapped to blow the magma apart. The material properties of a magma can be controlled by relatively small variations in its composition forged during magma genesis and the distinct journeys that different melts take from the mantle to a volcano's vent. For those living near an awakening volcano, whether it is likely to be safe to peer in close like at Masaya or whether considering a call to evacuate might be more prudent depends crucially on understanding this transit from deep to shallow. Taken to the extreme, the division between volcanic tourist spectacles and tragic disasters starts deep beneath the volcano and their preludes are written in the atomic-scale language of crystals, liquids, bonds and bubbles.

A volcano is the tip of a great unseen subterranean plumbing system. Fingers of magma reach up through Earth's crust from its mantle source. Sometimes I like to imagine magmatic plumbing as subterranean counterparts to the Earth's surface rivers, with deep catchments and tributaries coalescing into the immense reservoirs and channels that feed Earth's volcanoes. It's far from a perfect analogy but, as a river starts with the fall of countless raindrops far upstream, the rock that erupted out of Santa María in 1902 started as little globules of melt squeezed out from between the mantle's grains approximately 100 kilometres below the volcano. These liquid droplets are less dense than the surrounding rocks and so rise buoyantly. As raindrops merge and join to form first a trickle, then a stream, and eventually a mighty river, these melt drops gradually gather as they move, merging into larger channels pushing upwards towards the surface. While a magma is buoyant it will rise inside the Earth, first squeezing upwards seeking out the easiest way to stretch through the hotter, more elastic depths and then breaking or melting its way through the cooler, less yielding outer layers.

Where the Earth's crust is stretched and thin, like at mid-ocean ridges, molten rock might find its way quite directly to the surface. In thicker crust, like that forming continents, there is further to go and things are often more complex. While some pathways to the surface become well-worn, carving out efficient thoroughfares between deep and shallow, other routes turn out to be more roundabout. The more circuitous the route, the higher the chances that rising melts might hit dead ends, stalling as they run out of buoyancy or fail to break their way further upwards. Here they will pond, spreading out in horizontal layers – sills – perhaps cooling and freezing here forever or just staging on their journey before something pushes them on towards the surface.

Those first juicy drops squeezed from the Earth's mantle have a composition we call basalt. For a liquid rock basalt is quite runny and flows relatively freely. Sometimes these liquids find their way to the surface swiftly and have little time to evolve and change. The dark bulbous pillow basalts that push out onto the ocean floor at mid-ocean ridges have not had far to travel through the stretched Earth where the sea floor is splitting. Almost 70 per cent of the Earth's surface is formed of such basalt hidden as sea floor beneath the waves. But basalts can erupt on the continents too. At places like Masaya, the basaltic magma has found its way relatively directly to the surface with limited opportunity for its chemistry to alter, despite the thicker crust of the Central American isthmus. When a liquid basalt cools rapidly it tends to freeze into a dark and fine-grained rock, sometimes flecked with green olivine crystals stark against the black. It is largely this primitive rock, relatively fresh from mantle melting, that, given the right conditions, can cool to form the spectacular tessellated prisms, or cooling columns, found across the globe. These characteristic landscapes, like the great organ-pipe structures of Giant's Causeway in Northern Ireland or Fingal's Cave on the Isle of Staffa in the Inner Hebrides of Scotland, have impressed thinkers and inspired myths, legends and music celebrating their strange geometries and regularities.

Although Santa María's 1902 magma would have started as basalt down inside Earth's mantle depths, by the time it emerged it had changed. The magma that meted out such damage that dark October day was both stickier than basalt and charged with higher levels of dissolved water primed to make it blow. For me in Guatemala in 2009, relating the dusty white-grey 1902 ash to the same common ancestor as the dark glassy products from Masaya's lava lake was not a mystery. These samples were just some among many, taking their places in a scientific system cataloguing and explaining the diversity of volcanic products, or as they are often more broadly termed, igneous rocks, strewn across the planet. But like rivers and magma systems, scientific thinking has a catchment and creating a genetic system for such igneous rocks needed confluences of different observations and ideas. Agreeing which rocks were and were not formed by magmatism was a crucial early tributary feeding into our current understanding of the igneous world.

Rocks in Volcanoland can take a great array of forms: foamy white pumices from Vesuvius; tumbling dark dense lava flows from Masaya; dusty ash from Santiaguito, the speckled shimmer of white quartz, pinkish feldspars and black sheety micas in the granites of places like Dartmoor; and the basaltic cooling columns of Giant's Causeway, to name but a few. Distinguishing this igneous menagerie of colour, texture and form from other types of rock was, understandably, not immediately obvious to earlier observers. Tracing their lineages of formation to a common ancestor, to basalt squeezing from the mantle, was a challenge, especially in a time before long-distance travel was as relatively commonplace as it is today and before field photography opened up vistas beyond those drawn or directly seen. Before technologies like rail, motor vehicles and flight facilitated journeying, it was less usual for the same individuals to have seen both fresh volcanic rocks and ancient ones now remote from the theatres of volcanic action. Thinking was much more dictated by the landscapes accessible to the thinker.

In the early eighteenth century it was widely believed in Western philosophy that volcanism was an accidental or recent phenomenon on the planet. Water was seen as the key agent for geological change. In this narrative, woven in part from biblical ideas of the deluge, the origin story of crystalline rocks, like basalts and granites, was one of precipitation from a widespread ocean. In this version of the world, known as 'neptunism', these rocks crusted or settled out of the lapping waves something like crystals in salt pans. Gradually these ideas ebbed and observations convinced more and more thinkers of basalt's ubiquitous volcanic origin. Evidence came from poring over the rocks in locations like the Auvergne in France, Giant's Causeway and Fingal's Cave. Observers logged the proximity, and sometimes smooth progression within a single rock mass, between ancient basaltic lavas, the great fluted prisms of columnar basalts and other volcanic products, more similar to those already commonly known at active volcanoes themselves. Greater travel and communication began to join the dots between the products of active volcanism rumbling out in places like the Bay of Naples and ancient volcanic rocks found in other places across Europe, far away from present-day eruptions. Chronicles from ever wider travellers showed that the power, action and agency of magmatism was widespread rather than something accidental or unusual. By the 1830s natural philosophers like Karl von Leonhard in Germany and George Poulett Scrope in England had built on earlier work by James Hutton in Scotland to unify ancient with modern basalts and postulated that heat and volcanism were central to the development of our planet's landforms. (A school of thought known at the time as 'Plutonism'.)

Coarse-crystalled rocks like granite took longer to bring into the igneous fold. In the eighteenth century many natural philosophers viewed granite as being a primitive rock with its origins going all the way back to the creation of the Earth – thought to be much younger in those times. Careful observation of rocks in the landscape and new ways of studying them in detail began to challenge this thinking. In the late 1780s James Hutton carefully mapped some of the rock formations among the breathtaking scenery of his native Scotland in places like Edinburgh's Salisbury Crags, Glen Tilt, Galloway and the Isle of Arran with an aim to determine whether granite had 'been made to flow in the bowels of the Earth' rather than dating from the planet's origin.

Hutton's foray into Glen Tilt granite hunting, as a guest of the Duke of Atholl, in 1785 was both fruitful and comfortable. Just ten minutes on foot from Forest Lodge, the Duke's hunting-seat where he was 'entertained with the utmost hospitality and elegance' he found the 'most perfect evidence' that the local granite had been formed after the other surrounding rock and had been made to break its layers (or strata) and 'invade that country in a fluid state.' In these rocks, carved by the River Tilt, he found fingers of pink coarse-grained granite levering between the layers of grey rocks where they met, speaking to him of younger rocky granitic melts squeezing – intruding – into the older metamorphic and sedimentary country rocks. These intrusive textures pushed against ideas of granites as the earliest rocks precipitating from a watery solution somewhere inside or outside the Earth, but were not enough to clinch the deal in favour of their magmatic origin.

To resolve these debates, we needed to see rocks in a new way. Much of our understanding of volcanoes today comes from linking the microscopic scale with the human – and ultimately the planetary – scale. In the 1850s, Henry Sorby began this process of expanding how we see rocks by delving deeper into their details. Sorby's fellow Yorkshireman William Williamson had learnt the art of stone cutting, grinding and polishing from his jeweller-lapidary grandfather and put these skills to use to thinly slice things like bones, teeth and fossil wood for study. He became an acquaintance of Sorby's and gave him a lesson in making such slices. Given his geological interests, Sorby began to make slices of rocks in order to study them under the microscope. Seeing their components and structure in these new levels of detail allowed him to draw new conclusions about the origins of rocks like granite. By studying samples created by known processes and comparing them to unknown natural specimens he argued that, under the microscope, they possessed 'sufficiently characteristic structures to point out whether they were deposited from solution in water or crystallised from a mass in the state of igneous fusion' – in other words from a magma. Under the microscope, similarities between granites and known volcanic rocks, invisible to the naked eye, were revealed. Study of these very small-scale structures was to be a giant leap towards identifying granites as igneous.

Intuitively, an appropriate scale on which to study rock might seem to be that of the pebble, boulder, cliff face or landscape. For these reasons, back in 1858, Sorby was not convinced that his fellow geologists would adopt this new microscopic approach to the study of rocks. Although apparently met with initial scepticism concerning the propriety of examining mountains with microscopes, Sorby argued 'that there is no necessary connexion between the size of an object and the value of a fact, and that, though the objects I have described are minute, the conclusions to be derived from the facts are great.' His view has ultimately prevailed. This new technology of careful slicing and polishing rock, reframing rock samples as 'specimens' and submitting them to forensic examination in intimate microscale detail, opened up a new way of seeing the processes making our planet. These days geological departments across the world are full of shallow trays housing libraries of these microscope slides – thin sections – of rock. They are sometimes breathtakingly spectacular under magnification: crazy pavings of crystals, strange geometries, or smeared impressionist canvases, each one its own scattered tale of the rock's life and times. They often do indeed have large-scale implications. In geology it is sometimes our business to see a world, or perhaps sometimes even more, in a grain of sand.

Agreeing which rocks were and were not formed by volcanism was a major gathering of the headwaters of ideas that flow into our current framework for understanding volcanic processes. Joining this with a system to explain the origins of this diversity would be the next challenge. For nineteenth-century European natural historians, this discussion started with attempts to explain the distinction between dark basalts and the other lighter-coloured flows – named 'trachyte' by French priest and mineralogist René Haüy. Across Europe, these different hues of rock were often woven together into volcanic landscapes in places like Skye, the Auvergne, Bohemia and Italy, presenting a very obvious mystery to be solved. If this cornucopia of different types of igneous rock were related to each other as this geographical association suggested, this left the puzzle of what determined which type of rock a magma would freeze out and how a magma could change from yielding one type of rock to another. In the weird world of Volcanoland rocky magmas don't just flow and froth, they also evolve.

As early as the 1820s, Scrope had presented ideas of lava changing as it ascended. Although largely rejected as speculations, Scrope's ideas sowed the seeds for future thinking. In the 1840s, Charles Darwin published his characteristically careful observations of basalt flows on the Galapagos Islands during his voyage on HMS *Beagle*. Based on these field notes, he proposed ideas of chemical evolution of magmas via gravitational separation of crystals of different densities. US natural historian James Dwight Dana was the expedition geologist on an 1838–42 mission to explore the Pacific. Here he observed the relatively fluid and gentle basaltic eruptions of Kīlauea and concluded that buoyant rising liquid rock left behind crystals that had precipitated at higher pressures and temperatures as they rose, and therefore changed. Importantly he linked his ideas with the formation of coarse-grained rocks like granites via slow cooling of the magmatic Mother Lode.

In the 1850s, German thinkers like the chemist Robert Bunsen working in Iceland and Baron Ferdinand von

Richthofen working in the newly discovered volcanic terrains of the American West expanded these burgeoning theories, broadening the diversity of processes and rock types that they captured to include magma mixing, melting of the crust and links to deeper actions of change. Gradually these principles were married into knowledge of Earth's thermal history, internal structure and magma sources. Sophisticated as these pioneers were, it is striking how their different ideas tied to a particular experience of volcanism – active or ancient, lavas flowing or exploding. The volcanoes weren't only shaping their rocky products but also leaving the fingerprints of their different characters on the thinkers of the age and the theories they proposed to explain them. The streams of thinking in the scientific watershed were shaped by the volcanic terrains they drew from.

As often with the flow of the human history of ideas, there were still meanders to navigate to get to how we see things now. Over the ensuing decades, powerful insights pulled together to channel understanding from many sources, for example: widespread study of the chemistry and texture of igneous rocks, experimental studies of tiny crucibles of rocky liquids by innovators like Norman Bowen at the Geophysical Lab in Washington, DC and careful field studies of great exhumed layered magma bodies, like the Skaergaard Intrusion in Greenland. By the 1950s, the idea of magmatic refinement in the crust as a dominant process leading to the great diversity of igneous rocks patchworked across our planet was firmly embedded. Today it guides how we imagine the very nature of the continents and how we conceive the processes driving the character of all our world's volcanoes.

Magmas are stacked in great storeys below our planet's areas of active volcanism: vast ladders of sills and upwards intrusions linking us at the surface with the inner Earth. At every layer, melts are potentially refined, for as they cool the chaotic scramble of the liquid's molecules starts to slow and patches of order, that is to say crystals, emerge. These crystals pull certain atoms out of the liquid and into their structure and leave other chemical elements behind in the fluid's chaos. The lighter elements are, on average, more likely to get left in the liquid, making these melts less dense and more buoyant, allowing them to pull on upwards free of the heavier crystals. Shedding crystals inside the Earth changes the chemistry of a magma. In 1897 George Becker, one of the earliest geologists to join the US Geological Survey (USGS), likened this process of refinement by fractional crystallisation to common methods used to strengthen solutions in salt extraction or beverage production. 'A bottle of wine or a barrel of cider exposed to a low temperature deposits nearly pure ice on the walls, while a stronger liquor may be tapped from the center.' Importantly, in magmas, silicon and oxygen (known together by geologists as silica) and water are among the chemical components that tend to be focused into the remaining liquid. Magma distillation within the Earth brews ever more silica- and water-rich liquids: basalt becomes andesite then dacite then rhyolite – with each stage the rocky liquids generally conspire to become stickier and more charged with water, increasing their explosive potential.

But much of this magma will never see the light of day. It is hard to know how much stays within the Earth. Studies like those that I described in Chapter 1, showing the mismatch between Masaya's gas emissions and erupted magmas as well as many other whispers of evidence from the Earth's depths, yield some clues. It seems likely that at least as much magma (and maybe, in fact, many times more) gets left behind, deep and cryptic, as erupts. Hugged and insulated by the Earth's crust, these unseen abandoned melts cool slowly, creating coarse-crystalled, intrusive, rocks like granite (with similar overall chemistry to erupted rhyolite) and gabbro (with similar overall chemistry to basalt). The large crystal sizes in these plutonic rocks – named after the ruler of the underworld in Greek mythology – are testimony to their relatively slow cooling. This gentle ebbing of thermal energy gives the liquid's molecules sufficient time to self-organise and grow into the large crystals that yield the visibly grainy complex mosaics that make them such striking rocks as countertops or paving stones. In contrast, finer-grained or glassy rocks often tell of a history of faster cooling without time for large crystals, or maybe even any significant crystals, to grow although a rock's chemistry as well as its thermal journey also plays an important role.

As well as being beautiful rocks, granites can also create stunning landscapes. Most of the rocks of my beloved Dartmoor in Britain's south-west corner were once a slow-cooling plutonic liquid mush. In the forest by the River Plym under the Dewerstone cliffs, as in many places across this grand granite terrain, your feet crunch on a sand of rose-tinted feldspar and clear quartz crystals nearly 300 million years old. On a family outing I picked up a palmful of these grains to show the children and tried to conjure for them from their leaf-dappled glint in the summer sunlight the great batholithic magma body within which these crystals grew. I imagined what great temperatures, pressures and upheavals these crystals had been through to eventually sit catching the sun in my cupped hand by that babbling stream. It was hard to tell whether they were impressed.

But far away from Dartmoor, in places like Guatemala, it is the consequences when these more-evolved magmas do make it to the surface that are of very real and sometimes urgent importance. For the focusing of silica and water into the magmas as they journey through the crust alters their behaviour in fundamental ways, changing the very manner in which these liquids flow. Unlike liquid water, a molten silicate rock is not made up of simple tidy molecular units. In a quantum dance of their electrons, silicon and oxygen form larger structures, interlinked chains of silicon atoms bridged together by oxygen atoms. These complex twisting polymer networks snag on each other as they move, giving lava flows a strength and stickiness far greater than more familiar water-based – aqueous – fluids.

Basalts are already far stickier than water. At the surface their runniness – viscosity – is something like a thick wallpaper paste. As magmas evolve to higher silicon and oxygen concentrations, the molecular networks become ever more extensive and cumbersome and the oozing liquids become ever stickier – more viscous. Although when I imagine this polymerisation process I sometimes get an image of the thickening custard crust that I used to detest so much when compelled to eat it at school lunchtimes, in a magma this change is neither trivial nor a small effect. Evolved silica-rich magmas can be thousands to millions of times stickier than a basalt. And as their chemistry changes they become potentially more dangerous.

Deep within the Earth, water is dissolved in magmas. As the rock rises and the pressure drops, the magma cannot contain its water anymore and it escapes into bubbles of vaporous steam within the sticky liquid. It is like opening a soda bottle (or a champagne bottle if you prefer Lacroix's more decadent analogy) and watching it fizz; the bubbles dance out of the solution and escape in trails from the open neck. Although it is hard to discern by peering into a fizzy drink, these bubbles grow as they rise. This effect is easier to see if you lie at the bottom of a swimming pool and let out a breath – the bubbles expand a little as they pirouette up from you before bursting at the surface. The swimming pool water offers little resistance to their growth or motion.

Magmas are not as free and easy but a runnier basalt, with generally lower water content, has a far better chance of allowing its gas to bubble forth freely within the volcano or at the surface than its more evolved cousins like andesites or rhyolites. The bubbles bursting from a seething lava pit, like that I was fortunate enough to see at Masaya, do certainly bear at least some basic resemblance to a foaming ocean. Sometimes, a basaltic magma might be moving upwards too fast to get much of its gas out before reaching the surface, leading to the great fizz of a fire fountain, but these do not tend to have the sheer explosive power of eruptions like those we have explored from Vesuvius, Santa María or Pelée. These more ferocious eruptions are driven by the more refined, higher water content and far stickier andesites, dacites or rhyolites. In these more silica-rich magmas the bubbles can barely move and must fight to get free. Consequently, melts have a greater risk of arriving in the shallow crust without significant gas having been able to escape and with the pressure inside each bubble straining to break from the constrained matrix of the liquid rock. At some point the overpressure in these bubbles might reach catastrophic levels, to such an extent that it shatters the magma from the inside, driving massive explosions.

Big explosive volcanic eruptions like that at Santa María in 1902 unfold via a series of mechanical failures at the atomic scale. In the Earth's crust, a large body of magma, drawn from the volcano's deep catchment, had accumulated in its shallow plumbing, priming the system for this huge outburst. For things to begin, somehow the strength of the crust failed and the magma opened a crack to the surface. Once open, the higher pressure deeper within the Earth squeezed the magma up and out. When an explosive eruption gets going, the magma is moving very fast towards the surface. Like the bubbles rising to the surface in a swimming pool, the fast-dropping pressure during this race upwards means that the magma's steam-rich bubbles will seek to inhabit an ever-expanding volume. The changes are huge. Over the top five kilometres of the magma's journey, the pressure it experiences decreases by more than a thousand times. The volume that the bubbles would ideally inhabit increases by a similar factor. Eventually, the chemical bonds holding the molten rock together simply can't cope and around the largest bubbles the strength of the hot, crystal-bearing liquid fails and breaks apart, fragmenting the magma. This catastrophic mechanical failure creates a mixture of pumice blocks and ash flowing up with the hot gas. This releases the gas pressure, which had been contained by the liquid. It is now abruptly let loose, and its sudden expansion blasts the hot rock and gas out of the volcano's muzzle, sending shock waves through the air and ground: sound as a blow delivered by air, which caused our intrepid Eisen to need to clutch onto the hillside to avoid being thrown over.

The great human tragedies administered by the volcanoes of the Caribbean and Central America in 1902 took minutes to days to unfold but inside the Earth these events had been brewing for hundreds, maybe thousands of years – or even longer still depending on how deep you draw their individual roots. Volcanoes are the taps of vast dynamic systems of melt pockets, magma channels and mushy reservoirs that span the Earth's crust and reach into its mantle. The differences between the spectacular fizz of Masaya's lava lake and the great mountain-breaking explosion of Santa María start deep within the Earth, written in the different patterns of the great catchments of rock underwriting each volcano's story. It is by studying the rocks thrown out that we know all this; and subtle changes in the chemistry of their lavas can give us important clues about the unseeable current inner workings and likely future behaviour of volcanoes like Santa María's present-day protégée, Santiaguito.

On Santiaguito in 2009 we were on a hunt for rocks. Its regular explosions have made it a popular target for volcanologists. They bring with them high-tech monitoring equipment to measure the sound energy of its booming blasts, view its gases as they pulse through the dome's carapace and capture in slow motion the concentric rings that open to let the Earth's detonations through. Scientists want to know what secrets these signals contain, to help to predict the volcano's next move. The slow squeezing out of sticky lava to build lava domes like at Santiaguito is only possible because its slow rise from within the Earth allows time for the trapped bubbles of gas to escape without catastrophically shattering the magma as in 1902. This can happen despite the magma's high viscosity as during its very slow ascent and cooling in the five to ten kilometres below the surface, the magma becomes brittle and small microfractures open, releasing the pressure held in the trapped bubbles more gently. These gases find their way out through the fumaroles around the dome's architecture and cause the relatively small frequent explosions that I described earlier, and their continual release avoids a more prolonged and violent episode like that of 1902.

Lava chemistry is a key part of this story and a paper about Santiaguito in the early 2000s had noted a decrease in lava silica content since the 1970s and had linked this runnier, more mobile rock to a trend towards longer lengths of lava flows down the sides of the edifice. But these observations also posed new questions about how the volcano's explosiveness and the frequency of its activity might change into the future. More systematic work seemed needed, but first we needed to get some rocks from the latest flows to analyse.

We studied the maps in great detail while planning the expedition. A colleague with a wealth of local wisdom had recently created a beautifully updated and detailed geological map chronicling the history of the volcano and tracing the outlines of its lava flows. On the map, these flows blossomed like petals on the page, coloured from purple, for old, to red, for young, from 1922 to the most recent (at that time a 2006 flow). On one version he had even helpfully shaded in an appropriate red the 'don't even think about it' zone, the immediate fallout area from the plumes and pyroclastic flows from Caliente, which we should avoid entering at all costs. I shuddered watching

YouTube videos filmed by adrenalin junkies on the volcano's domes. We consulted other scientists who'd worked there before and decided that securing samples should be possible, if maybe a little challenging.

Understanding a landscape on paper is very different to understanding it in reality. This is true in general but even more so in volcanic landscapes. You can read endless scientific papers about a place and pore over every map and photo available but there are always essentials that evade you. The first time I visited Mount Saint Helens in the USA, almost four decades after its devastating and much written about eruption in 1980, I felt the awe of meeting an idol and making the happy discovery that they were even more fascinating and magnificent in real life. The photos and maps in papers and books could not do justice to the scale of the landscape, the great amphitheatre scar in the mountainside and the graveyard of tree trunks still left from the destruction. Other times the reality of volcanic terrains can rapidly lead to the realisation that what looked like a straightforward hike on the map is actually a gruelling battle through dense vegetation, up loose slopes or over blocky lava fields. Arriving at Santiaguito we quickly realised that we would be facing all three of these testing terrains and our assessment of acquiring samples as 'a little challenging' began to feel like something of an understatement.

On that visit the Santiaguito Volcano Observatory was up a rough road through land belonging to one of the local coffee plantations. We would sometimes travel up behind some of the trucks bringing in the workers on our way there. The volcano's coastwards flanks are still dominated by these *fincas* resonating with the descriptions and photographs captured by Eisen and Anderson over a century earlier. These days Guatemala is no longer remote and globalisation has reached its tendrils into the livelihoods of its citizens. Signs at some of the *finca* entrances boast participation in Starbucks coffee production. The beans ground to make many an urban morning latte might have been dusted by volcanic ash on these slopes.

Sitting in the observatory, creepers and fern-like fronds framed the view of lush jungle with the mountain presiding splendidly in the background; it seemed to me like a paradise. But a landscape is the layering of stories and history, both geological and human. Later, reading accounts by the Guatemalan activist Rigoberta Menchú, who was awarded the 1992 Nobel Peace Prize for her social justice work promoting the rights of the indigenous peoples of Latin America, I dwelt upon the much darker narrative layers hidden among the the forests, coffee crops, mountains and brooding volcanoes of the country's landscape.

Inside the simple grey block of the observatory building was a special window and a desk on which sat a pair of binoculars used by the observers who watch the volcano, providing radio reports to Guatemala City. The observatory had its own small rock collection but it was far too precious for us to take back to Oxford to chop up and powder. With the observers' help we were there to get our own specimens. Our first sampling target was a cluster of different lava flows from the 1980s and 2000s – nestling into the valley of the Nima II River. Armed with packs full of water, food, geological hammers and Ziploc plastic sample bags, we set off from the observatory to drive as close as possible to where these tongues of rock stretched away from the volcano.

One of the observers, Julio Cornejo, came as our expert guide. The volcano rose above us ever larger as our four-wheel drive bumped along the rough track, until the road ran out and it was time to pile out and continue on foot. The jungle was dense and heading into it was like taking a plunge into a dappled green and earthy ocean. It also felt unwise to lose sight of the volcano grumbling above me. Julio seemed unfazed. He set off, wielding his machete to cut a path for us, and we descended into a densely vegetated gorge, slipping down the steeper parts and scraping out musky tracks behind us. The air was moist and creepers and ferns brushed my arms and face as I pushed my way through. Leaves, like giant hands, blocked out the sunlight and where its rays stole between them it picked out dusty shards in the soily air. Every now and again an ominous distant booming reminded us that the volcano was not far away.

Eventually we dropped out of the jungle into the deep drainage gully of the Nima II River. In the rainy season these clefts flood with water and loose pyroclasts from the volcano, sometimes gushing down onto the villages below. Everything was coated in fine flour-like ash. The broader leaves captured speckled patterns where wind or rain had brushed or washed the ash off. We scaled the other side of the gully and I remember, despite the dry weather, being relieved to be out of the path of potential flash floods. Now we were finally onto the blocky lava flows themselves. Some of these flows had only been created a few years prior: light grey, with smooth faces and sharp edges. Mosses and small ferns were just starting to colonise them in their more sheltered nooks. Up close, fine crystals were visible on the less weathered surfaces. Occasionally we could trace an irregular clot of even finer grains, weathered deeper into the rock face: a sign of something complex and enigmatic deep below.

The volcano loomed over us, puffing and rumbling – like some giant silicic dragon sitting on its spoils. Maybe it was the sense of being in the dangerous presence of something far more powerful than myself, but J. R. R. Tolkien's illustration 'Conversation with Smaug' – in which the tiny invisible hobbit, Bilbo Baggins, speaks with the immense dragon curled around his gleaming treasure heap – sprang to mind. The lava flows presented new challenges to navigate. The work of the day, lurching around and hammering off samples, was punctuated by the volcano's threats. Plumes curled into the sky above our heads and, more alarmingly, trails of hot rock and gas – pyroclastic flows – would descend its side towards us before disappearing into a ravine between us and the mountain. Each time I held my breath in fear that the cloud would make it over the hummock protecting us, but the 'don't even think about it' zone had been drawn on sound principles and we were never in real danger. Between explosions white dust devils sometimes hovered over the volcano's summit – tell-tale wisps of gas pushing dust out from inside the domes without it exploding. The volcano was everywhere. Its rocks were under my feet and grazed my hands and elbows as I scrambled. Its ash was in my hair and coated my face, neck and limbs. My ears were tuned to its every rumble. It was a mesmerising and intense place to be.

As the afternoon's clouds started to gather, we loaded up our rocky cargo – meticulously bagged and labelled with GPS locations and the name of the lava flow it came from – and set off to hike back to the four-wheel drive. Getting up the steep jungle ravine with tired, battered, dusty bodies and packs full of rocks was a far greater challenge than coming down. Rarely used muscles screamed as I pulled myself up from root to root, dirt coating my grazed hands and elbows, smearing brown dust over my grey ash tarnish, soil now blending with the ash in my eyes and hair. Julio made it look easy but the rest of us panted and cursed. I wish that I remembered more about the wildlife that we saw that day in the jungle. But I do remember vividly when one of the team reached for a root above them only for it to slither away and out of their grasp just in time. We were cautious of using the treacherous trees to help us climb after that and later convinced ourselves, with the help of a faltering internet connection, that it was most likely a kingsnake, and definitely not the highly venomous coral snake that also apparently lurks in the forests.

A few weeks after our return to the UK, our samples arrived in Oxford in packing-size cardboard boxes. Although of little commercial value, these boxes stuffed full of pieces of the Earth's crust are the starting point for many volcanological studies. We prepared them in much the same way that Sorby had in the 1850s, selecting and slicing samples into thin sections. Under the petrological microscope these revealed what crystals the rocks had borne with them on their journey from the Earth's insides. Many of these minerals held clues about the inner workings of the volcano. Spongy textures on the outsides of some of the crystals told of how they started to re-dissolve in the liquid magma as it rose. The thickness of these porous rims allowed us to estimate how quickly the magma had travelled upwards during eruption. With other clues in the rocks we aimed to shed light on Santiaguito's future destiny, like whether it will carry on with its small explosions or if there is any risk of it starting to build up to another 1902-scale cataclysm. The suite of rocks and data that we gathered in 2009 suggested its lavas had continued their trend of reducing silica, therefore getting generally less sticky. We therefore deemed it possible that its next flows might continue to extend longer distances from the volcano into the surrounding tropical forest. But it also means that Santiaguito is not showing signs that it is likely to threaten the people around it with a major eruption in the immediate future. This work suggested that over the decades the volcano is gradually working its way down through the layers of its magma supplies to the less distilled, less evolved melts slightly deeper within its plumbing. The rocks gave some assurance of relative, compared to 1902 at least, safety.

But overlaid on this apparent relative stability, the chemistry and physics of Santiaguito's rocks remain finely poised between the benign and the threatening. In 2014, some years after we were there, part of the volcano's dome collapsed into the Nima I River and a series of pyroclastic flows threatened the observatory. Julio filmed this and I watched it intently in my university office on YouTube. Tempest Anderson and his colleague John Flett described the billowing clouds of the pyroclastic flows from Pelée after the 1902 disaster as 'cauliflower convolutions of dust and ash'. 'Cauliflower' captures well the shape but seems an entirely too homely likeness for these menacing clouds and in Julio's video they are more like some horror-film monster. As they reached towards him they pulled down great trees with a sickening crunching sound. After this event the observatory had to be moved further away from the volcano and into a different *finca*.

Then, in 2015 and 2016, despite silicon and oxygen in the magmas dropping still further, the volcano spent two years exploding more violently, sending plumes over five kilometres into the sky and hollowing out Caliente's previously flat-topped dome. The quiet microfracturing of the ascending magmas, key to the calmer release of the volcano's gases, had for some reason shut down to such an extent that the trapped bubbles were being carried upwards with greater overpressure and eventually pulverising the hot rock in more violent explosions. These stronger outbursts scattered ash, and sometimes larger-sized fragments – lapilli – almost daily over local communities up to 30 kilometres away.

Since 2016, at the time of writing at least, Santiaguito has settled again and returned to quieter activity more like when we were there. A new dome has grown to infill the crater at Caliente's summit and the new lava flows that were pushed out between 2014 and 2015 into the canyon of Nima I's headwaters are already being incised by the river's action. Some small shift in the inner physics and chemistry of the mountain's plumbing transforming again the mood of the landscape and its impacts on local people.

But there are other ways, beyond the immediate threats of explosions, flows and ash, in which a volcano can lash out against the local people, and these restless mountains do not only act alone. Changes in the hazard that a volcano represents to the local population are mediated by other external agents in the natural environment, as well as those from within the Earth. Today, as in 1902, when the loose volcanic pyroclasts and blocks are subjected to Guatemala's periods of extreme rainfall, new problems unfold. Another day in 2009, still sore after the hack through the jungle, we visited the area downstream from the steep slopes of Santiaguito. Here the tight gullies of the Nima I and Nima II tumble straight down from the volcano's flanks and merge to form the Nima River. Where the rivers widen or turn they shed their load, leaving silt and boulders in their wake. We picked our way among these boulders and across the rickety wire and plank bridges spanning the rivers' rough channels. Some twisted uneasily as we crossed them and their lurching groans and the generous gaps between planks made me more than a little dizzy when I looked down. I am not sure whether I would qualify for Eisen as a 'not too timid traveller'.

The town of Old El Palmar nestles around a tight 'S' shape in the River Nima. When we visited the river seemed small and innocent, sticking to the middle of its sandy channel with room to spare each side, despite the ominous size of many of the boulders strewn along its banks. But, although the town is cradled by the river, it has been far from nurturing to the small community. When heavy rains fall on the volcano upstream, they mix with the constant accumulation of volcanic dust and rock to form a terrible slurry, Eisen's 'thick gruel-like mud'. These lahars frequently threatened to engulf the small settlement, with the muddy waters rising to get around the river's tight bends and eventually the village was overrun by floods of mud from the volcano in the late 1990s. Most of the people have resettled in New El Palmar, founded a few kilometres away. In 2009 some of Old El Palmar's ruins overhung the river channel. The overgrown concrete, metal and breeze-block skeletons of former dwellings jutted out into the channel of the river. The triangular peaks of a roofless gable, rising defiant and angular over the rubbly banks, were festooned with creepers and ferns. The abandoned homes in which families once cooked, ate and slept, will one day topple into the river completely.

Like many other volcanoes, there are local legends around the mountain's activity. In the area around El Palmar one tale is of the volcano's owner, the supernatural Juan Noj, who has his house inside the mountain. Since it is always erupting, this house is constantly burning down and Juan Noj must send disaster and pestilence to the local towns, or bribe individuals with wealth in order to harvest souls to help him rebuild it. One version of the story has a group of neighbours from the old town of El Palmar hearing the sound of Juan Noj's horse's hooves one midnight and going to seek him in the River Nima's narrow valley. Juan Noj appeared to them as a tall, blond character with eyes the colour of fire mounted on a beautiful white horse. The horse seemed to trot on a blanket of white smoke without touching the rocks in the river's ravine. The neighbours made a deal with him: he would give them wealth and prosperity but would return within 30 years to take their souls to do his work within the volcano. Juan Noj then disappeared, the sound of horse's hooves echoing around the men and the singing of owls chasing them as they stumbled home in the pale moonlight.

The town prospered as promised, but the debt was not forgotten and as the date of payment approached the neighbours enlisted the help of a Catholic priest. The priest promised to rid the town of its curse and its debt. Juan Noj took revenge by first sending a dense column of ash to the town's outskirts, followed by a torrent of mud and rocks. This deluge passed through the centre of the Catholic Church and completely destroyed the town. Although local understanding of the volcano has moved on from Juan Noj, there is a sense of potent agency in the ruins and rocks strewn around Old El Palmar and the continuing activity at Santiaguito even if the deeper power is from the Earth itself rather than a troubled apparition.

Supernatural forces aside, the intersections between the human and volcanic spheres are complex. Activity at Santa María makes it an ongoing threat, always there in the background of the lives of those around it. For experts like Gustavo, intent on protecting his country's people from harm, constantly active volcanoes like Santa María and Fuego present a specific set of challenges. While there is an ongoing supply of fresh data and rock samples from these busy mountains, when you can safely reach them, their continuous activity can make it hard to predict when the threat rises, even just a little, but enough for the volcano to lash out, as Fuego did with fatal consequences in 2018. However, the constant presence of flows and explosions makes the local population alert to the dangers and keeps settlements away from at least the very most hazardous locations. We are still working to understand the general traits of systems like Santiaguito, but it seems likely that to fuel another outburst the size of the 1902 eruption, the edifice would first need to seal and fall quiet before accumulating sufficient magma in its shallow plumbing.

During this more dormant time, memories of the mountain's volcanic nature might fade and it would be the job of Gustavo and his colleagues to maintain vigil for signs of reawakening and try to keep a suitable awareness of the danger alive in the population. They have experience of this, with at least 30 potentially active volcanoes tracing Guatamala's Pacific rim, most of them quiet throughout written history. Some of these volcanoes are known to have produced much larger eruptions than that from Santa María in 1902. Managing these volcanoes presents a different set of challenges from those of Santiaguito or Fuego. It asks us questions about how we reconstruct patterns of volcanic behaviour where human accounts are absent and the system is at rest, about how we understand the hazards that these volcanoes present when left only with the patchy remnants contained within a landscape's rock record. The answer is that when we cannot read the account of a volcano's past in human words, we can at least learn to read the partial accounts hidden in the rocks. It is a skill that must be learnt, and it is where our adventures will take us next.

Chapter 3

Reading the Rocks

*How do we understand past eruptions
and predict future volcanic hazards?*

I have already described how working on volcanoes can sometimes bring them to life, as if they have their own characters. Maybe this anthropomorphism draws from a similar vein to the myths and legends that swirl around volcanoes like Juan Noj at Santiaguito, but it is not unusual for human folklore to hold elements of truth, and there is at least some scientific resonance encapsulated within these framings. As human characters often slip into patterns of behaviour, frequently the past is the best key we have available to understanding the future of a volcano. Magma pathways through the mantle and crust can remain similar over thousands of years and more, and the eruption styles that are governed by the movement and chemistries of these melts can often also repeat themselves, at least to some degree. Many volcanoes settle into their own patterns of behaviour, and if we can untangle their key traits, they can give us powerful clues about what will happen next. For eruptions, like those of Santa María, La Soufrière on St Vincent and Pelée on Martinique in 1902, we have written and photographic accounts of how these relatively recent cataclysms played out. But what do we do when key information about a volcanic system goes deep beyond the human historical record?

On the geological timescale, human history is brief and, given the age of many volcanoes, the need to cast our eye deeper into the past is very often necessary even for those with recent major eruptions. How else are we to answer all-important questions about their wider dispositions, such as how often we should expect them to produce the largest events in their repertoire? Delving deeper into the long timescale of Volcanoland is also necessary as Earth has not shown us all its tricks. We know that the planet is capable of much larger scale eruptions than those that we have directly chronicled. These rare but extreme events will almost certainly occur again and so we must be on our guard; we must consider where, when and how to recognise them and what predictions we might make about how they will play out. But can we really understand them when all we have left are the rocky products of these vast eruptions that time has spared? To reconstruct these past cataclysms, we must read the rocks left behind. It is one of the key skills for any field volcanologist and one I first learnt in the idyllic Aegean, on the band of islands known collectively as Santorini. This archipelago numbers a mere handful out of Greece's thousands of islands but here, within these picturesque fragments of land, lurks the evidence of one of the largest eruptions globally in the last 10,000 years. It's all there, written in the landscape, once you know how to decipher it.

From above, Santorini's biggest island, Thera, looks something like a green and buff crescent moon against the deep blue of the Aegean Sea. Broaden your gaze a little to include its island sisters and you will see that with the curling wisp of Therasia and the tiny, pale splinter of Aspronisi, Thera makes a shattered ring and that these fragments of land enclose a deep lagoon. Sharpen your focus and you can see that far from being a perfect crescent, Thera's shape is more a series of scallops or bite-marks – nested calderas – where the waves of the enclosed sea cut at the land hollowing out the island's belly. But the lagoon is not just filled with the lapping sea. Two darker shapes blossom above the waves in its heart – the archipelago's newest islands: Palea and Nea Kameni, old and new 'burnt' islands. The larger Nea Kameni in particular seems to radiate out from its centre in subtle petal-like lobes breaking the glistening waters.

Although it looks nothing like the archetypal conical volcano, Santorini is a very romantic place. The lagoon is edged with sheer cliffs which tumble down several hundred metres into the Aegean. The sharp escarpments of these cliffs are striking, not only for their precipitous gradient but also the vivid and varied rocky layers that they present to the bay. These stripes of bright or tanned ochre, rusted red and dark grey strata are dusted in green where vegetation has managed to find a foothold. Towns frost the clifftops with white houses. Among them is Oia on Thera's northern tip and the island's main town – Fira – which nestles on the rim in the crescent's navel. On a sunny day the azure domes of the islands' churches chromatically echo the sky. Dark paved streets and steps – some flagged with chiselled lava – creep down the cliffs where they can. Hotels advertise cave rooms hollowed out of the cliff face and caldera-view infin-ity pools. Brochures boast the most beautiful sunsets in Greece and tourist bars promise the best local wines to sip while soak-ing them in.

The view over the lagoon is a picture postcard. On dark and dreary days elsewhere, I can conjure it to my mind. The alluring Aegean sparkles in the sun, punctuated by the white puffs of waves and lacing smudges of boat wakes. In this classical image of bright island paradise, the dark blots of the 'burnt' Kameni islands may seem out of place, but in the summer, the regular stream of tourist boats buzzing to and from these islands and the invitation to relax in the hot springs around their shallows is as enticing to many as the rest of the scene. But you would be right to read something more ominous in the presence of the Kamenis and their barren outlines. Deep beneath this tranquil tourist playground, the Earth is anything but peaceful. The tectonics of this region are complex but deep beneath the Mediterranean waves to the south of Santorini (beyond Crete) the African plate pushes north and subducts, yielding an arc of volcanoes from Methana, just tethered to the Greek mainland, to the island of Nisyros, just shy of Turkey. Santorini is one of the most active volcanoes in this Hellenic Volcanic arc and everywhere in this stunning scenery are clues of past and future tumult. The spectacular cliffs with their intricate layers are a voluminous history book and once you learn to read the rocks, thousands of years of the islands' story return your gaze, a tale stretching back far beyond antiquity.

Overlain upon these deeper happenings is the human story. Santorini has hosted human civilisation for over 4,000 years, chronicled in the islands' archaeology as well as some written accounts. The modern-day life of the islands is tiered over the remnants of these past inhabitants as well as the geology – layer upon layer of human as well as volcanic stratigraphy. And in the human and geological now of Santorini the volcano's inner world underlies the island's inhabitants and visitors whether they realise it or not. Often unnoticed, the trappings that translate these subterranean machinations permeate the island's people too: over recent decades, threaded within the thrum of human daily busyness, networks of scientific sensors strain to capture any signs of shifts in the volcano's subterranean state. Seismometers monitor the tiny earthquakes that herald the movement of magma or gas as well as the rupture and creaking of the region's straining tectonic plates. Permanent and precise GPS receivers record the changing shape of the land over time. This meeting of timescales – from the hundreds of millennia of the rock record to the period of human history, decadal scientific monitoring and sensor signals available in almost real-time – is a powerful lens for expanding our understanding of volcano science. This layering of information is one of the things that makes Santorini such a special place to work. Pulling these strands together allows us to see much deeper into the volcano's inner character. It allows us to recreate the lost landscapes and happenings of the volcano's past and, perhaps more importantly, it aids our predictions about what might happen next.

Over its long and rich history Santorini has been known by many names. The ancient Greek historian Herodotus records Stronghili (circular) and Kalliste (fair one) as early monikers. Thera derives from the Spartan commander who founded a colony on the island in the ninth century BC. Santorini is a more modern Venetian name. It honours Santa Irini (Saint Irene), martyred in Thessaloniki in 304 AD, and is a reference to a chapel erected on the islands in her celebration. But perhaps Santorini's most famous historical association is with the Bronze-Age Minoan culture. Centred on the island of Crete more than 100 kilometres to the south, and named after the legendary King Minos, the Minoan civilisation is entwined with Greek myths of Theseus, the labyrinth and the Minotaur. But it was also the first civilisation in Europe to leave behind evidence of large multi-storey buildings, artworks, tools and writing. Santorini was an important location for the Minoans and the massive explosive eruption some 3,600 years ago, known as the Late-Bronze-Age eruption but also sometimes referred to simply as the Minoan eruption, has drawn the strong association between the islands and this lost civilisation into the modern volcanological lexicon of the islands.

Geologically speaking, the fallout from the Minoan period

is quite literally everywhere on the islands. Looking south from Fira town, along the island's tumbling cliffs, there is a yellow layer of rock about 30 metres deep that cloaks the clifftop and underpins the houses and roads. Although it is not visible everywhere, burrow beneath the topsoil and we find that this light-coloured mantling mass is spread across most of the land of the archipelago. In some places it is quarried out of the cliffs for use in construction. Working quarries are ideal places to get a closer look if you can get permission to safely access them. Up close, this great smooth sandy stripe is a broken mass of fragments of – now wind-scoured – pumice and ash that were thrown out of the volcano. Even without delving into any of its rich detail, the thickness of the layer, especially when you stand at its foot and it looms above you, gives a sense of the scale of the event.

Looking more closely there are many clues about the violent story told by this ochre outcrop. Stretched and angular pumices, pulled easily from the cliff face, tell of frantic motion. Dense boulders the size of large watermelons sag into the stratigraphy: volcanic 'bombs' blasted out of the volcano. The deep lagoon itself speaks volumes about the destruction that this eruption reaped on the islands. The force of the explosion destroyed not only the land that hosted it but also hollowed out the Earth from the inside, so that it collapsed back in on itself and formed the most recent of Santorini's calderas. Such was the force of the eruption that it committed the site of its own vent to the depths. It is only by piecing together the evidence we find in the cliffs, quarries, shoreline and sea floor of Santorini today that we can do the most complete job reassembling the events of this enormous eruption and conjure the lost form and shape of the islands before their map was so violently redrawn.

It was not only the physical backdrop of Santorini as a volcano that the eruption redefined so utterly, but also a landscape that was home for the islands' Minoan inhabitants. As well as the story of the physical eruption that is hidden in these rocks, they also capture the remains of a very human tale. During the construction of the Suez Canal in the second half of the nineteenth century, large quantities of pumice were quarried out of this upper layer of Santorini and exported to build the waterway's concrete piers. This mining uncovered buried structures on both Therasia and Thera and excavations in the 1860s led to the discovery of a significant Minoan settlement near the present-day village of Akrotiri on the southern limb of the main island. Among those intrigued by these discoveries was the French geologist Ferdinand Fouqué, whose 1879 book on the islands, *Santorini and its Eruptions*, was to become a standard reference for geologists and archaeologists alike. Excavations are ongoing and have reclaimed a wealth of important Minoan artefacts from the volcanic rock. Vividly preserved wall paintings depicting Minoan culture, now mostly displayed in museums on Thera and in Athens, are particularly captivating: flowers, birds, boats, boys boxing – a society that had time to play as well as work. The large, complex ships depicted on the walls of the Akrotiri excavations tell of the importance of trade to the Minoans and it seems somewhat apt that the opening of an important new artery of commerce at Suez led to these new discoveries about the Minoans' past.

Despite this intimate entwining of an eruption and a people via the hidden ruins within the physical fabric left by the volcanic tumult and the geological designation of this violent episode as the 'Minoan eruption', questions regarding the actual proximity of humans to the full onslaught of volcanic fury remain. No bodies have so far been found within Akrotiri's remains and few trappings of wealth like jewellery or gold or silver. The site seems to have been largely abandoned before its burial in rock and ash. Unlike at later-interred Pompeii, it seems that the Bronze-Age population of Santorini had sufficient warning of the severity of the volcano's intentions to trigger their escape and with sufficient time to pack at least their most valuable possessions. These warnings or 'precursors' likely included an earthquake sufficiently severe to cleave stone staircases, crumble walls and collapse roofs, and early explosions from the volcano, already large enough to scatter significant ash over the town. What became of Akrotiri's inhabitants remains a mystery. Some archaeologists speculate that they may have been killed at sea or in encampments on the plains outside the town. There is sketchy evidence that some may have fled to neighbouring islands. Whatever their fate, the archaeological digs at Akrotiri have not told of horrific last human moments like the mass graves of Pompeii and Herculaneum do.

Although the local population in Akrotiri seems to have escaped immediate doom, the slow decline of the Minoan civilisation at about the same time has led some to speculate that the eruption might have sparked a shift in regional power. Ash fall damaging agriculture on Crete or volcano-induced tsunamis that destroyed ships and compromised Minoan naval power have both been fingered as possible triggers. Drawing direct causal links this far back in history remains elusive. Nonetheless, the eruption has assumed a strong, if far from solid, mythology of upheaval. Among other narratives of disjuncture, writers have sought to link it with Plato's story of Atlantis or the biblical Exodus of the Israelites from Egypt. There is little or no evidence to substantiate these claims, but Santorini sits in an area of the world with a long history and tales that we still tell today were unfolding not far from the volcano at the time of the eruption. Perhaps the temptation to use clear evidence of a volcanic cataclysm to make sense of important events and narratives lost in the mists of time is unsurprising.

Whatever the wider consequences of the Minoan eruption, the impact on the islands themselves was profound. According to Herodotus they were not inhabited again until five generations before the Trojan Wars, that is to say about 1330 BC or

250 years after the eruption. Not only were the islands engulfed in falls and flows of volcanic rock, the entire morphology of the islands was changed with the combined forces of explosion, collapse and flood. Unlike more recent eruptions, such as Santa María in 1902 or even Vesuvius in AD 79, we have no written accounts of what happened over those fateful weeks in the Bronze Age and it is not even clear that humans were present to witness it up close. In fact, we lack good accounts of a caldera-forming eruption of this scale globally over the whole timespan of human experience. There is much that is unclear about how they unfold.

Eruptions of this size are fortunately rare, likely only one or two per 1,000 years, but their consequences, especially on today's crowded and interconnected planet, could be grave and far-reaching. Their rarity, while lucky for human society, adds to the challenge of understanding what will happen when another such event does inevitably occur. Despite its antiquity, the Minoan eruption is, in fact, one of the most recent examples of activity on such a scale, making it a key focus for volcanic study. What's more, the great scalloped cliffs of the flooded caldera slice through time for us, showing off the strata from the volcano's older eruptions, which might otherwise be buried deep and inaccessible. The same colourful rocky layers that capture the Sun's evening rays so spectacularly and delight the tourists, also enthral volcanologists with the lessons they reveal about the islands' long past.

Over the years, geologists have wandered Santorini's islands and carefully described, sampled and analysed the different rocky stripes that make up its stratigraphy. This collective effort has pieced together not just the Minoan eruption, but many other volcanic events in Santorini's past – big explosive events like those in the Bronze Age, and smaller ones too – allowing scientists to look for patterns in the volcano's behaviour. The clues are not just in the rocks themselves. Other lessons can be learnt from the shape of the islands' land and the sea floor. Gradually, over the centuries, explorers and scientists have reassembled the anatomy of the islands' past destructive cataclysms, as well as their renewal. When the time comes and the volcano again erupts, we hope this collective reading of the rocks will have armed us or our successors with the knowledge to recognise what sort of activity to expect and to predict what might happen next. Time will tell.

Santorini is an incredible place to learn field volcanology. I first went there in 2001 to study such skills just before starting my PhD. I have returned many times since, sometimes now as the years have turned and student has become teacher, to lead undergraduate training trips myself. I led such a trip while heavily pregnant with my first child – and fully admit that I had grossly underestimated the discomfort of long, hot days carrying my precious load, and the difficulty of finding comfortable maternity 'field work' attire. Every time I visit I see new things, but I will never forget the intensity of that first outing. We had come as part of a package holiday group to keep costs low and took buses around the island that were often crammed and sweaty in the late summer heat, despite the first tempering edge of autumn in the breeze. We stayed on the flatter side of the island away from the caldera in a town called Kamari. Here the land rolls gently into the sea and the beach has dark volcanic sand. Cool dips in the Aegean were a very welcome end to days of dusty field work. It was just after the attacks of 9/11 in the US and the world felt ruptured and edgy even in this distant and idyllic location.

A key aim for our expedition was to work through the field evidence of how the Minoan eruption progressed from its start to its finish, and early in the trip I found myself face to face with the imposing pale ochre cliff of Bronze Age rock in a quarry just south of Fira perched on the caldera's sheer rim. I had come to

Students studying the Minoan deposits, 2022

volcanology from an undergraduate training in chemistry and, excited as I was to start this new chapter, finding the meaning in this towering rocky canvas was daunting. The pale-coloured rock face stretched up tens of metres above me and seemed at first glance both to exist as an imposing uniform bulk and in intricate and overwhelming detail. Where to start and how to

make sense of it? Often on field work we are pointing sophisticated spectrometers at the volcano or taking samples back for
complex chemical analyses in the lab. However, when reading
the rocks to decipher the process of past eruptions the first
stage is usually simply to sketch and describe what you see and
uses nothing more sophisticated than your eyes, a notebook
and a pencil.

There is not just one way to see and describe a snatch of landscape like the cliff of the Minoan eruption. Since childhood I
have enjoyed playing with paints to capture the countryside.
When art is the aim, you can choose which things to focus on
and which to ignore. If I had been there to paint the Minoan
deposit that day, I might have stood back and concentrated on
the blocks of colour. I might, for example, have reduced the
scene to the broad undulating shape of the bright bulk of the
ochre sandwiched between the blue of the sky and green of the
scrubby vegetation. For a volcanologist, too, it is important to
take a step back and trace these broad outlines into your field
notebook. But we try to be more formulaic with our record and
field notes include a checklist of things less likely to be found
in an artist's sketch pad: details of your precise location, the
compass direction of your outlook, estimates of the angles that
the rock makes with the horizontal and some sort of scale bar,
for instance. These days we would usually take a photo of the
most important sites as well to complement our sketching. For
geologists each detail might hold an important clue. The subtle
changes in texture up the cliff face must not be ignored. Your
sketch must capture the speckled detail of the lowest part of the
cliff, the fine and smooth laminations in the layer above and the
more massive, chunky quality to the uppermost parts dappled
in the shade cast by the shallow hollows scoured out by the
Aegean's salty weather.

I have never been very proud of my field sketches, but certainly some geologists would make very fine artists. Similarly,

many landscape artists seem to have an almost intuitive sense of the rock they paint. Many years later at an exhibition in London, Georgia O'Keeffe's incredible paintings of New Mexico unexpectedly called my mind back to sketching in Fira quarry. Although O'Keeffe was gazing upon the rocks and cliffs for very different reasons from mine when viewing the Minoan deposits, there was something deeply resonant in the colours and textures she had evoked in her pale volcanic rocks against the blue of the New Mexican sky. It was nice to reflect that this intimate act of looking at a landscape in detail to capture it is something that artists and geologists share.

As well as being part of the place's beauty, each subtle layer in the Minoan cliff has meaning. Once we had taken in the cliff as a whole, it was time to get closer. A warm wind whipped up dust, stinging my eyes as we approached the rock face, donning hard hats in case of falling debris. Close-up, the speckled texture of the lower cliffs resolves as piece upon piece of volcanic rock so loosely held together that you can easily pull chunks out with your fingers. Most of this layer is made up of light-coloured, pebble-sized pumice pieces (clasts), many of them stretched and angular. But there are also darker bulbous fragments dotted among them and heavier, denser rock chunks as well. I followed the base of this layer along, as it coated and mirrored the undulations of the old surface below. This cloaking – or mantling as we call it – of the old land is a clue that these rocks fell from above rather than flowed into place. The fact that they still crumble out in your hand, rather than being welded together, suggests they were relatively cool by the time they fell here and no longer hot and malleable enough to press together and merge. Their broad similarity in size is another clue that they have been through some grand sorting process. Sifting through all this evidence of falling and cooling we can rewind the eruption and imagine this layer of broken clasts falling as a rain of pumice pebbles from the Bronze-Age sky. It tells us of an immense eruption column rising from the vent like 'an umbrella pine', to use Pliny's words penned over 1,000 years later, with smudged trails of dusty debris falling from its trunk and bows.

It is unlikely that we will ever know what humans saw of the Minoan eruption. But we might imagine the people of Akrotiri casting their gaze back towards their former home as their ship beat against the waves in escape. This first phase of the eruption, which generated the pumice fallout in Fira quarry, would have resulted in the ominous rise of a pillar of billowing debris – a Plinian column – punching up into the sky as if its rock weighed nothing, and spreading high above. We know from the distribution of ash layers found in places like Kos, Rhodes and in cores of sediments from the Mediterranean Sea floor that the cloud spread to the south and east, driven by winds high in the atmosphere. While it would depend how far you were from the volcano whether day would have been fully turned to night, if you were a Minoan survivor making for Crete, it might have felt as if the dark blot of the eruption swelling in the heavens sought to chase you as you fled.

Like the other big explosive eruptions we have met in Volcanoland so far, the roots of the volcanic paroxysm on Bronze-Age Santorini lie deep within the Earth. We know from the style of the eruption and the rocks it produced that the magma was rich in silicon and oxygen and therefore sticky and viscous; pressurised and primed to detonate. Although the Minoan eruption was many times larger than the 1902 tragedy at Santa María in Guatemala, there are fundamental similarities in the processes that generated the Plinian phases of these two events. As alluded to in Chapter 2, the details of how and when an eruption plays out are often written in the shallow plumbing and surface situation of the system. The rapid rise of sticky magma packed with trapped overpressured bubbles sets the scene for violence. As the magma rises and the bubbles enlarge, the strength of the sticky liquid will eventually fail.

Suddenly released, the gas is free to expand unbounded into the atmosphere above, unleashing the full impulsive force of the explosion. Rocks are fired out at speeds of hundreds of metres per second. Great shock waves spread out through the air and the ground. The Earth quakes and breaks.

With the release of all this raw kinetic energy of violent movement, it is tempting to imagine that the swathes of pumice and denser rocks I sifted through in Fira quarry were simply fired there from the volcano's mouth. We might envision their tracks like bullets from a gun arching in a ballistic parabola from the vent to fall to rest in layer upon layer on the islands' clifftops. But studying the fallout patterns from eruptions like the Minoan shows that things cannot be this simple. Powerful as they are, these ballistic forces alone cannot get ash and pumice as high into the atmosphere or as far away from the volcano as it often gets flung. Ash layers left from the Minoan cataclysm can be traced as far away as the Black Sea, telling us that the eruption column must have towered up to 35 kilometres above the volcano, up into the Earth's stratosphere. As we saw before, the direct human observations from Guatemala in 1902, chronicled by Eisen and Anderson, measured Santa María's debris ascending to a height of 30 kilometres. This is higher than a bullet would reach if fired upwards from ground level with a plausible volcanic ejection speed – or 'muzzle velocity' to use the language of ballistics – as estimated from measurements and calculations. The kinetic energy from the explosion becomes exhausted only part way up the trunk of the eruption column and the rocks would fall out too close to the source to explain the deposits we see. At first glance, perhaps, these mystic rocks might even seem to defy gravity!

But while volcanic eruptions shatter vast volumes of rock, they certainly do not break the laws of physics. It is just that the force of the explosion at the base of Pliny's 'umbrella pine' of hot rock and gas is only part of the physics that drives this mixture to rise to such lofty altitudes. It is only relatively recently that volcanologists pieced together a detailed explanation for what Pliny so evocatively recorded all those centuries ago. Until the later part of the twentieth century, the pyroclastic rocks, like ash and pumice, produced by these violent fragmentation processes, were often overlooked in favour of studying denser volcanic rocks like lavas. In my much-thumbed copy of his classic 1993 textbook *Volcanoes: A Planetary Perspective*, Peter Francis attributes this neglect to a certain geological snobbery about these very friable field deposits. In his words 'it is hard to give an ash deposit a satisfying swipe with a hammer to get a hand specimen of proper rock'. Ground-breaking (pun intended) work in the 1970s by George Walker of Imperial College, London and Robert Smith of the USGS, among others, began to systematically describe the deposits left by very explosive eruptions like the Minoan and then to explain the physics of their formation.

Simply put, it is not solely the speed or kinetic energy that drives an eruption column upwards. Experiments and simple mathematical models of gas and particle behaviour in the second half of the twentieth century showed the importance of the heat or thermal energy guarded within each of the eruption's rocky clasts. This heat, brought up from inside the Earth, can drive convection, carrying the eruption plume to far higher altitudes than would be possible with its kinetic energy alone. But hot rock on its own will not become buoyant compared to Earth's gassy atmosphere; the heat energy within the solid clasts must transfer to the surrounding air, reducing its density and giving it lift. To do this, the volcanic column must first mix the cooler surrounding air into its mass, folding it in around its edges as it blasts out from the vent and continues upwards. This air gets heated up by the hot rock fragments, causing it to expand and become buoyant compared to the air outside the plume. As the plume travels further up, more and more air mixes in and warms, keeping it buoyant and driving it higher and higher.

Earlier, we imagined the rocky clasts as a hail of bullets fired from the volcano's mouth, but once a Plinian column lifts off, a new analogy is, in fact, more apt. Violent as the column is, it is as if the hail of eruption debris transforms into countless miniature gas burners firing the rise of a huge fleet of tiny hot air balloons. It is this billowing process of hot air rising by convection that grows the chunky trunk of a Plinian eruption column into the sky. Destructive as the consequences are, the physics of a volcanic eruption always forms a mixed metaphor in my mind: an arsenal of ballistic force feeding a dirty cloud of billions of sky lanterns. Nonetheless, it is a cloud that can darken even the brightest day and rain down devastation.

But what of Pliny's branches and the distinctive spreading umbrella shape of the top of such plumes? Although the rocks coming out of the volcano's muzzle are very hot (many hundreds of degrees Celsius), their heat energy has its limits. Once their warmth is exhausted they can no longer drive convection and the plume will reach a level where its density equals that of the surrounding air. Carried by its momentum, the plume overshoots this point of neutral buoyancy, and then falls and spreads under its own weight. High up in the atmosphere the plume is like a fountain, fanning outwards as it falls like spray from the nozzle of a hose. Like a fountain, in a strong breeze the wind can also play a role in determining its shape, pushing more of the dusty flow along the axis of its blowing. But unlike a water fountain, the plume's material will not fall all the way to the ground – not immediately, at least. For as the mixture of gas and rocks cascades downwards through the sky's storeys, the thickness or density of our planet's atmosphere increases. Eventually the still-warm plume will again find its neutral buoyancy level, where its density matches that of the surrounding air, and here it will spread like a tree canopy.

Throughout this convective rise and fountaining spread, the plume will shed chunks of rock sorted by size: the largest nearest the vent and the smaller fragments ever further down its reach. It is from the trunk and branches of the Minoan eruption cloud that the sharp pebble-sized pumices fell into what is now the Fira quarry. I have a couple of the stretched and glassy chunks that I collected during my 2001 visit resting on the shelf above my office desk today. Sometimes among the many demands of the everyday, I try to catch a moment to imagine their violent birth, their furious rise and their final fall, followed by thousands of years lying peacefully hidden and undisturbed within

Stromatolite (lower image) and pumices, from the author's collection (pen for scale)

the landscape until plucked out by my younger self. There is a lot to be read into these rocks.

But this broken layer of up-and-down-again pumice is not the end of the drama written in the quarry wall. Above the rubble of the fall deposit there are further upper seams, or units as we sometimes call different strata like these. I picked out two clear higher units above the pumice fallout easily, even when sketching the cliff for the first time. The thinner middle unit is fashioned from smooth laminated sheets of fine dusty ash. The thicker layer above – about ten metres thick here in the quarry but up to 35 metres elsewhere on the islands – is more massive: a jumble of blocks of different sizes held in a matrix of fine pale ash with characteristic hollowed weathering. Glassy black fragments glint in this upper layer, punctuating the white pyroclasts and telling, as we will come to later, of the islands before the eruption.

The sense of movement written into these upper layers is very different from the lowest tier of fallen pumice. Rather than mantling the old land as the rocks rained from above, these units infill any ancient undulations, levelling out previous topography. These features, and the muddled sizes of the clasts in the uppermost layer, tell of emplacement here by flow rather than falling. Something had changed in the eruption, something fundamental, meaning that the plume could no longer loft and rise but instead surged over the islands, leaving thick blankets of rock cover in its wake. Understanding how and why this change occurred was the next challenge in our field training.

To understand what was going on during the eruption, we need to reimagine the geography of the islands. It is not straightforward but by roaming, describing and sampling Santorini's terrain, geologists have sought to wind back the clock to see the place as it was as home to the Minoan traders of Akrotiri. Surveying the cliffs beneath Fira town from a boat bobbing in the bay, your eye might be drawn to the white houses perched on the edge, or the zigzag path tracing the sweaty ascent (unless you give in to a donkey ride or the cable car) from the Old Port quayside up the imposing topography, or the many brash stripes of the caldera wall's rocky units. You would be forgiven for missing a pale smudge of dusty rock clinging to the cliff face just over halfway up, like the last vestiges of snow hanging on in a hollow after the thaw. As subtle as this pale patch might be, it tells a tale. Closer analysis shows it is made of the same type of pumice I pulled out of the fall deposit in the quarry on the cliffs high above. This pale little remnant suggests that, rather than being part of the collapse associated with the Minoan eruption, this section of cliff was there when the first Plinian stage of the Bronze-Age cataclysm unfolded and that these crags captured some of the pumice falling from the plume. This dates this sheer face as the wall from a much earlier caldera collapse. These cliffs, which are now a backdrop to the coming and going of modern-day tourist and fishing boats, were also part of the scenery against which Minoan life played. A shared view, a tie point in time between now and then, despite the eruption's huge upheaval.

But while this section of cliff was not significantly changed by the Minoan eruption, much else about the islands was. More clues come from the rocks thrown out by the eruption. Back in 2001, another day took us to the Minoan deposits, near the northern town of Oia. Like Fira, Oia perches on the cliffs on the caldera's rim. Below the town is its port and beach and it was among this scene of seaside idyll that we had paused to lunch on the freshest squid I have ever eaten, fried by the fishermen on a camping stove straight off the boat. While it had been tempting to linger in the bay we tore ourselves away and back up the hot hill to the less salubrious location of an old quarry, being used as a rubbish dump. Here most of the Minoan deposits had been hollowed out for building materials, but among the useful pumice the miners also found an abundance of denser rocks they could not use and left behind in the quarry, conveniently accessible for us to study, until more recently when the site has been developed as a hotel. Picking around the abandoned white goods and sofas and through the volcanic debris here, the field leader plucked a chip of one of these blocks and handed it to me. The chip felt heavy in my hand compared to the lightness of pumice. Rather than a complex web of bubbles frozen in the act of escape, this rock's complexity came from the layers of grey, white and pink interwoven in crenulated blooms, like tiny petals. This was a fragment of sedimentary rock whose origin had nothing to do with the igneous expression of the volcano, but its presence gives us another important clue about the form that the volcano held all those millennia ago.

This little section of rock curling into my palm was a chip of stromatolite. This was not a rock type that I had heard of before and I felt somewhat cheated, having worked hard to acquire some literacy of the igneous world, to be faced with even wider rock diversity at this early stage of my studies. Unlike the frenetic instant that froze the form of the erupted pumice blocks, the layers in the stromatolite fragment were built up gradually over time. The colourful, complex layers magnified under my hand lens were laid down by colonies of microscopic organisms, probably cyanobacteria. Although plentiful in some parts of the fossil record, stromatolites are rare on the present-day planet, their minute microbial builders depleted by the grazing of more recently evolved animals. Where they are found, in places like Shark Bay in Western Australia, they characterise shallow, brackish water environments. On Santorini, however, my little chip, which I still have in my rock cabinet today along with my pumice fragments, is not rare in the Bronze-Age debris. Over a thousand stromatolite blocks have been counted cluttering the pale Minoan pyroclasts. Their presence tells us that rather than today's deep bay, the people of Akrotiri would have paddled in a shallow, salty lagoon within an old caldera pinched into the island's northern sector. The rocks, including the hummocky stromatolites of this ancient inlet, were ripped asunder during the melee of the eruption's violence and thrown out among the rest of the debris.

This information about the pre-eruption geography of the islands helps us to read what's laid out in the ochre upper layers above the well-organised pumice fall deposit back at Fira quarry and to trace the anatomy of the eruption. The size-grading of the pumice fallout locates the eruption vent near or within the shallow lagoon. The middle unit of delicately laminated sheets of fine sandy ash tells of energetic explosions with the pulverised rock thoroughly diluted by hot gas. Flash boiling of external water when it laps into a vent can do this. Way back in those frantic hours of the Bronze-Age, the brackish lagoonal waters likely found their way into the tumult. Each incursion would have detonated like a bomb blowing steam-rich dusty blasts sideways out from the plume's trunk, flowing over the land to deposit the fine layers of ash that we still see today above the pumice fallout.

Eventually the explosive force of the water flooding the vent was strong enough to blow large chunks of the plume off sideways. The heat energy taken from the hot pyroclasts to boil the invading water likely quenched the plume's energy to rise too. The sheer mass of rock forcing its way through the ever-widening vent at higher and higher rates also set the buoyant forces of convection a mounting challenge in terms of countering gravity's pull back towards the Earth. Acting together, these processes disrupted the column's convective rise. No longer buoyant, it collapsed under its own weight to flow with ferocious force over the ancient islands, coating them in the massive thick, blocky upper layer in the cliff that loomed above me in the quarry. With the centuries of accumulated knowledge, sitting sketching at the quarry cliff face conjures the days of ferocious eruptive process that have written themselves into the scenery to face the millennia.

The deposits show us that eruption is a story of water as well as searing rock and, although not as easily captured when visiting Santorini, the seabed as well as the land is marked by its action. At some stage during the eruption the old lagoon boiled dry and was likely infilled and blocked off from the sea by the volcanic debris. More and more magma poured out from the volcano's mouth until eventually the hollowing out of its innards caused the overlying Earth to give way and collapse, pumping out more magma as it sank. This great sagging deepened the old caldera to the north and stretched it to the south to create the distinctive deep bay that dominates the present-day landscape. It is hard to imagine while watching waves lap the islands' shore today, but evidence suggests that this gargantuan caldera, some ten kilometres long at its widest point, was initially dry after the cataclysm. It would have briefly formed a huge chasm, lined with the sandy Minoan pyroclasts and likely still erupting. At some stage, likely not very long after the sinking of the caldera floor, the wall of rock keeping out the waves failed allowing the sea to deluge into the crater. Evidence from the seabed still tells of where this first forceful flow occurred with bathymetric maps showing a huge scalloped amphitheatre-like structure just outside the bay's northern entrance, lurking beneath the Aegean waves. Today tourists flock to photograph the view from Oia towards Therasia where once, for a couple of days, a mighty waterfall flowed infilling the bay through the knick of this northern breach. While the link between the eruption and Atlantis is in all likelihood entirely spurious, such dramatic flooding seems more than worthy of this legend.

Back on our imagined Minoan ship packed with Akrotiri's escaping refugees, these ongoing phases of the eruption would have brought new terror to the view back towards their former home. The seawater incursions into the vent would have rung out booming explosions with rings of debris blasting out from the base of the eruption cloud. Then as the eruption column's buoyancy failed, they would have watched as the massive dirty morass of the volcano's plume fell from the heavens to engulf the islands themselves. Perhaps this was the last view they ever saw of their island home. Did tsunamis chase them, triggered as the volcanic debris poured into the sea? Did they feel the Earth shudder as the caldera sank? We are unlikely to ever know. They would have been powerless, with little choice but to turn their back to the destruction and grieve. A scene of such violent devastation seems hard to imagine gazing over the idyllic bay today. It is something I try to contemplate though as I sit sipping my drink and watching the evening sun redden the impressive golden layer capping the present-day cliffs.

Humans like patterns, and Santorini's cliffs are full of them on all different scales: the leopard spots of wind-weathered scallops, the downward scars and fans of landslides, the upwards cut of exposed dykes standing proud through the softer strata and layers of stripes on the scale of the cliff face and layers within layers. The deeper you look the more complexity there is, but if you stand back there is also some order. The Minoan deposits are the easiest layer to spot in their prominent place at the top of the cliff sequence and, as we have seen, evidence of the eruption's destructive power lurks everywhere in the island's landscape. But this is not the only time in the islands' long history that they have blown themselves apart.

Distinctive as the golden Minoan crest to the cliffs is, its features are not unique in Santorini's caldera walls. Looking north from the modern-day port of Athiniós on Santorini towards Fira town, about a third of the way down the nearest headland – Cape Alonaki – you can spot another sandy-golden, wind-scalloped layer facing the waves like a sunken, squeezed facsimile of the Minoan.

This layer, less romantically named Lower Pumice 2 (one of the earlier names for the Minoan was the Upper Pumice), is evidence of a previous eruption of similar scale and style to that in the Bronze Age but over 150,000 years earlier. I drew this cliff section on my first trip to Santorini and I was delighted to later find the same section sketched and reproduced in Fouqué's 1879 book on the islands. For hundreds of years scientists have been reading in the rocks the story of this little piece of Earth that seems doomed to make and break itself over the millennia.

With careful mapping, logging and sampling of Santorini's cliffs and its other rocks, geologists have put together hundreds of thousands of years of its volcanic history. Through this work they have traced two cycles of explosive activity at the islands, each one lasting about 180,000 years. Both cycles were formed of a sequence of at least five large explosive eruptions, each 20–40,000 years apart. The size of the eruptions and the silica content, and hence viscosity, of the magma broadly increases as we rise through the strata laid down by each of the two cycles: meaning the eruptions got more devastating as the sequence progressed. Many of these eruptions sank calderas like the

Santorini caldera cliffs at Cape Alonaki looking north with Fira town in the background. Visible in the cliff face are layers of darker lava and paler pumice including the deposits from the Minoan eruption (top) and the Lower Pumice 2 eruption (the next pale layer down).

Minoan. Evidence like the dusting of Minoan pumice clinging on to the old caldera wall below Fira suggests at least four of these huge collapses have occurred, bite marks defining the islands' outlines. The oldest collapse so far identified followed the powerful Lower Pumice 2 eruption, the culmination of Santorini's so-called first eruptive cycle. Far from being a unique cataclysm, it turns out that the Minoan eruption is rather the most recent episode in Santorini's pattern of supremely self-destructive behaviour stretching back way beyond antiquity.

Interleaved with these records of past destruction, however, we also see something else. Between these highly explosive episodes, the island has rebuilt itself and again there is much evidence of this rebirth written into the landscape once you know how to read it. There in that Cape Alonaki headland that Fouqué sketched all those years ago, sandwiched between two layers of reddish-ochre explosive pyroclasts, are the dark blocky forms of thick lava flows. Even in the destruction of the Minoan, the abundant glassy black fragments I had spotted in the upper layer of massive pyroclastic flow deposits in Fira quarry are evidence of a substantial volcanic edifice growing up among the stromatolite boulders in the shallow salty lagoon and blasted apart by the eruption. Perhaps the most dramatic view of volcanic rebuilding is at the northern tip of Fira town in the towering headland of Cape Skaros. Here, hundreds of metres above the glistening waters of the Aegean, stacks of dark frozen lava flows are piled upon each other, jutting out over the sea. A blocky plug at the headland's crown echoes the funnels of the cruise liners that frequent the bay and almost gives the impression that the whole promontory is a ship ploughing into the waves. It is, in fact, the remnants of a great volcanic edifice that defined the islands 50,000 years ago or more.

We do not even need to read the ancient cliffs to see evidence of volcanic rebirth. The dark blossoms of the low Kameni islands in the bay have built up, layer upon layer, from the floor of the immense flooded crater left by the Minoan eruption. Like their predecessors, whose remnants now form Cape Skaros or the Alonaki lavas, these islands are gradually growing to fill in the void. They build up during lava-dominated eruptions, much smaller than those that have left the pale pyroclastics in the cliffs. Eventually their bulking mass will somehow reset and prime the system to brew the next large-scale explosion. Studying the Kameni islands we can decode many messages of hope about the much smaller-scale activity characterising the volcano's recent behaviour. But hidden within their dark dacite lavas and murmurings of unrest are sombre warnings that Santorini is not done yet and one day, likely in many thousands or tens of thousands of years, it will revert to its old mega-destructive ways.

Shifting our focus from the towering cliff faces to the dark, barren rubbly islands within Santorini's lagoon brings us swiftly into the geological present. If you narrow your eyes so that you exclude the caldera walls, viewing only the craggy outlines of

The Kameni islands with Therassia in the background from Fira town, 2022

the Kamenis, your time horizons have pulled in from the hundreds of thousands of years in the dizzying succession of the cliffs to viewing a mere 2,000 years – you step from geological time into human history. The youngest rocks date back to just 1950, squeezed out in one of the islands' shortest known eruptions. Unsurprisingly, perhaps, given that they are the site of all the most recent eruptions, for many years the Kameni islands were the main focus of attention for those studying Santorini's volcanic activity. It was the 1866–70 eruption of Nea Kameni that attracted the attention of Fouqué, who was dispatched to Santorini by the French Academy of Sciences in the first year of this volcanic episode, the first of several visits. Fouqué's 1879 monograph starts with the rich accounts of the first emergence of Palea Kameni above the waves millennia before. The Greek thinker Strabo and the Roman philosopher Seneca each recorded vivid details, likely drawn from the writings of the earlier Greek philosopher, Posidonius. Although both were writing over a century later, clues date these events to around 197 BC. Strabo is more precise on the location of the new land, but Seneca's account is perhaps more evocative, telling of how the sea foamed and smoke rose from its depths before the 'fire' became visible and 'from time to time, it expanded like flashes of lightning ... Then stones were thrown out wildly ... In the end, one could see protruding from the highest point a peak that had a burnt appearance'. This 'burnt' peak continued to grow, forming the nucleus of the humped profile of Palea Kameni that we see today, rarely visited by tourists and apparently home only to a herd of goats and a solitary hermit.

Although with some swirls of uncertainty deeper back in time, historical accounts of the eruptions stretch from these beginnings right up to the present. Many of these reports include beautiful sketches that depict the new flows and domes, and they allow us to trace the pieces of the episodic volcanic jigsaw that has assembled today's island-scape. The final eruptions building Palea Kameni broke forth in AD 726 and then after several centuries of quiet, in the 1570s the locus of activity shifted north-east and the newer island, now known as Nea Kameni, started to form. Fouqué draws accounts of these eruptions from many, often ecclesiastical, sources. Athanasius Kircher, the same monk whose vivid cross section of the Earth's inner fire I admired in Chapter 1, records the birth of the earliest part of the new island recounted to him some years later by the Jesuit missionary Father Richard. This first companion to Palea Kameni was the little dome-like cone of Mikra Kameni, now absorbed into the mêlée of Nea Kameni's interlocking flows. Nonetheless, its 'deep cavity with a rounded funnel-like form', once its own miniature island in the bay, is still visible from the right angle, mysteriously rather smaller than you might expect from the accounts that the 'volcanic fires lasted a full year'.

Perhaps my favourite story from Fouqué's wonderful compendium is an account of the 1707–11 eruption found in the archives of the Jesuit mission of Santorini. It is attributed to a Father Gorée and tells of a mass that seemed to be floating in the water near Mikra Kameni, first taken for a shipwreck, but soon recognised as a new bank 'formed of black rocks with white ground in the center [sic]'. These turned out to be blocks of porous pumice with 'the color and texture of loaves of bread'. Those who landed could collect fresh oysters and sea urchins from the rocks until suddenly they 'felt the ground shake and the reef began to tilt'. The sea seemed to boil and turned yellow, giving off 'suffocating sulfurous odors', and dead fish floated to the surface. 'The island grew before their eyes; in a few moments it rose seven meters and spread laterally to twice its former diameter.' Over the next few years, in fits and starts of small explosions and rocky flows, the eruption built about a square kilometre of new land with a characteristic lava flow that still stretches out into the sea from Nea Kameni's north-western corner like an accusing finger.

There is barely any shade on the Kameni islands, and in summer the temperatures here can soar. Tourist boats run from the main island and there is a well-worn path up to skirt the volcanic craters at the low summit, a short stroll for day trippers before a relaxing dip in the tepid springs that rust the water in a shallow creek formed between two lava flows in Palea Kameni's sheltered eastern shore. Away from the sparse paths, progress around Nea Kameni's terrain requires clambering between sharp crate-sized lava blocks, similar to those of Santiaguito. Like Santiaguito, these flows are blocks of sticky, viscous lavas, with their higher silica content making flowing a far more tortuous exercise than for basalts like at Etna or Masaya. In places you can see rubbly evidence of rock falls from the steep leading edge of the advancing lava – the flow front – underlying the massive blocks of the main flow. This tells of a motion more like a caterpillar track, crawling over the earth, rather than a more liquid-like stream. Patches of grass and lupins cling on in this otherwise hostile environment.

Despite the heat and beating sun, Nea Kameni is a superb place to study and teach volcanic processes. Even from the tourist path you can learn lots about the island's flows, domes and fumaroles and I have spent many hot but happy hours tramping over the island's dark topography. Of course, not everyone sees quite this same degree of fascination in these barren islands as volcanologists. I remember on that first trip in 2001, the bafflement of our package holiday company rep at the hotel welcome drinks ('But it's just a big pile of rocks?') when we asked how we could manage to spend a whole day on Nea Kameni. She had clearly not met many geologists before.

Although many of the early artists did an impressive job, an aerial view is a big advantage when it comes to capturing a complex composite structure like Nea Kameni. Fortunately, we have been conducting aerial surveillance of these islands for over 75 years. We can tile together the monochrome photographs taken by British reconnaissance aircraft from the early 1940s and mark the subtle differences in shape and shade that trace the flows. But it is in colour images that you can really pick out the petals of the different volcanic episodes with their subtle differences in hues and texture and in 2004 and 2012 a UK research aircraft flew to and fro above the Kamenis, taking a mosaic of such photos. These same flights probed the islands in even more intimate detail too: scanning the terrain using pulsed laser light – LiDAR – to trace its height in a similar way to how sonar bathymetry from a ship uses sound to survey the shape of the seabed.

From this data we were able to make a digital elevation map – a DEM – of the shape of the land poking out of the waves. With five height measurements per square metre, this map of topography is in significant detail. But there is something more valuable in this data than simply capturing the land's shape in greater intricacy. Inside the computer we can zoom in and fly through the land to see it in new ways. We can cut cross sections through flows and measure the height of the banks or levées containing them. We can run lines down their lengths to trace their folds and ripples and pull these things together to learn about the way in which they flowed many decades or centuries ago.

Incredible as these volcanic islands are, what we can see is just the tip of the iceberg and much is hidden below the waves. After the overflights in 2012, we worked with Greek colleagues to stitch together our LiDAR map of the land with their sonar map of the sea floor. Contour models of Santorini and its depths had been made before, sometimes out of beautifully crafted wooden layers, but through the magic of pulsed sound and light we were able to drain the waters of the bay inside the computer and reveal the full majesty of the Kamenis in new detail.

The bay is around 400 metres deep in the northern basin, and from these depths Nea Kameni rises to just over 100 metres above the waves. The whole edifice takes up a volume of just over three cubic kilometres. This means that since the Minoan eruption it must have grown, in fits and starts, on average by just under a million cubic metres, or about 300–400 Olympic swimming pools of lava each year. On Santorini we can check this growth rate number calculated from our high-tech methods against the historical accounts from antiquity. According to Strabo and Seneca it was about 1,200 years between the catastrophic Bronze-Age eruption and the appearance of the first 'burnt' island. If we simplify the edifice as a cone, then a volume 400 metres high, in other words high enough to break the waves, with a radius on the sea floor of a few kilometres, could certainly be plausibly assembled at the average rates calculated above over the time period that the history books describe. Inferences based on our high-tech interrogation of Earth's present-day shape and the tale of the Kamenis from ancient human testimony coincide.

This marrying of the land's shape above and below the waves revealed other secrets about the volcano's past and future too. When we see the whole picture – both submarine and subaerial – the surprisingly small land-based size of the year-long 1570s eruption, which I discussed above, is immediately explained by the substantial fan of material plunging unseen down into the caldera's depths. With the magic of our computerised three-dimensional DEM we can also take the islands apart, in digital space at least, by masking out the flows in reverse order and getting the computer to make a best guess at the surface that was there before. Once the chunk of each flow is removed, we can run the computer's algorithms to add up everything between their virtual bumps and curves to estimate their individual volumes.

The rich written history of the place and its catalogue of eruptions, including that compiled by Fouqué, combined with their volumes extracted digitally from our DEM, means that we can look for relationships between the different episodes and their timing too. From this we discover that if we plot a graph of the number of years since the last eruption – the pre-eruption interval – against either the volume or the duration of the eruption that followed, we get a roughly straight line. This gives us a tool to foretell the likely progress of a future event. More data would be ideal, but we can follow the line of this graph to read off the predicted volume and activity period for the 70-odd years since the last small eruption in 1950. Our pragmatic prediction is that, should an eruption start soon (and I will discuss shortly what signs we might see warning of this), it would last two to three years and put about 0.1 cubic kilometres of lava onto the surface of our planet, most likely as a new layer or lobe of Nea Kameni.

These predictions are wrought from the marriage of human observations chronicled by Fouqué and others with the high-tech power of laser beams, sound pulses and computer analyses; a framing of the quantitative by narrative. From the solemn, bearded photo of Fouqué in the frontispiece of the 1998 translation of his book it is hard to imagine what he would have made of us still mining his treatise to predict Santorini's future almost 150 years later. Reading of his life in France spanning the 1848 revolution, the Franco-Prussian War and the Paris Commune, I find it hard to believe that much would surprise him. Would he be disappointed or impressed by our progress towards understanding Santorini's many enigmas? Either way I like to think that he would be pleased that so many volcanologists today still share his passion for this little ring of islands guarding their nook of the Aegean and caught in a seemingly perpetual cycle of construction and destruction since deep beyond human history.

In the same way that the view of the Kameni islands across the bay from Thera today captures just the tip of the much larger unseen volcanic edifice rising up from the caldera floor, our present-day understanding of Santorini's volcanic system is built on centuries of human observation and study. This background to the place is often absorbed into our implicit underlying knowledge rather than appreciated in its own right. How those of us that study its activity today see the island's landscape encapsulates all that we understand about its history – both human and geological – and informs the way we think about Santorini's future too.

As we have already touched upon, the relatively short time since the last eruption in 1950 means that we are almost certain that the volcano is not on a path to a new Minoan-scale event anytime soon – we think that building this volume of magma in the shallow crust takes very much longer. But by 'soon' I mean not within the next several thousand years, which is a far briefer period if we measure it against the geological timescale rather than the yardstick of human lifetimes. Mapping the recent eruption vents using our LiDAR and bathymetry data, suggests that the next eruption will most likely be on or near the Kameni islands, somewhere along a rather mysterious line of seeming structural weakness, sometimes referred to as the Kameni line, running through the caldera and following the trend of these two barren islands from the south-west to the north-east.

But as well as vital lessons from the past, the way in which we sense the islands in the 'now' has changed over recent decades in important and transformative ways. How we garner knowledge of the volcano has moved from being something more gradual in the study of rock or landscape to something very immediate and current. Today, unnoticed by most, sensor networks permeate the modern-day life and tourist bustle of the islands. Through these feelers, local scientists can sense the islands' movements and vibrations in a way that was not possible down through the layers of their long documented human occupation or indeed even the last eruption in 1950. Monitoring the volcano in near real-time will not only give more warning of its next reawakening, but it also gives us new insights into the inner workings of this sleeping giant.

In the autumn of 2010, one of our graduate students in Oxford started looking at satellite radar data from Santorini. The plan was to use the radar data to map out a small area of subsidence across the southern half of Nea Kameni. But then in early 2011, the volcano's mood started to show a distinct and troubling shift. The pattern of earthquakes in the area began to change. They were still too small to be felt, but the network of seismometers buzzed as the inside of the caldera started to host a series of small shakes with the dots of their epicentres marked on the map building up over the weeks, like a slow swarm of bees, to roughly trace the Kameni line.

The shape of the archipelago also started to change subtly. Data from the islands' GPS receivers showed an abrupt switch from holding relatively steady positions to a pattern charac-teristic of a gentle swelling centred in the heart of the lagoon's northern basin. Scientists, like me, watched the data come in as the northern end of the Kameni islands and Cape Skaros jut-ting into the lagoon north of Fira town started to rise out of the sea gently and imperceptibly, at a rate of just a few millimetres every month. It was as if some powerful subterranean being had started inflating a balloon in January 2011, four and a half kilometres down within Earth's crust. The authorities convened a Special Scientific Committee for the Monitoring of Santorini Volcano. Some of the earthquakes were large enough to be felt locally and caused some alarm. All the centuries of past study came into play to advise the local authorities what sort of erup-tion to expect and what the hazards would be for the people living within the volcano's close reach.

Thankfully, after about 18 months of subtle shaking and swelling, all went calm and fears of an imminent eruption began to subside. Without the concern of a looming crisis and the very necessary focus on when or if it was going to erupt, we were able to reflect more on what the data from this episode told us about how a large volcanic system like Santorini works in general. In fact, this short-lived swelling was a fascinating insight into how the volcano's plumbing system draws magma from the deep into its shallow crustal magma storage zone. From the measured bulging of the islands we were able to estimate that, by spring 2012, the final volume of the metaphorical underground inflating balloon driving the unrest was 0.01 to 0.02 cubic kilometres (about 4,000–8,000 Olympic swimming pools). Assuming this volume was largely magma, rather than gas, we noted that this was about 10–20 per cent of the predicted volume of the next eruption, based on collating past behaviour as described above. This told us something new: rather than recharging the volcano's shallow store via a gradual drip, drip of magma from below, the deeper reaches of Santorini's plumbing appear to push pulses upwards in much larger bursts, each one a significant proportion of a future eruption. Although some quakes were large enough to be felt, this pulse of material rising from our planet's depths was largely stealthy and, prior to our modern monitoring techniques, similar episodes might have occurred on the islands throughout their history and not previously been noticed. The progress of technology is opening up new ways to see ever deeper into volcanic processes, but in places like Santorini these novel methods are so much mightier when combined with observations often wielding nothing more sophisticated than a pen or pencil.

Although the 2011 unrest fortunately came to nothing, it reminds us that the volcano is far from dead and it is a matter of 'when' not 'if' it will erupt in the future. Having spent considerable time imagining Santorini's past, this activity gave me a chance to imagine its future. The course of the next eruption, which based on past behaviour we might expect sometime in the next century, will most probably follow the historical accounts of those from Nea Kameni. It will likely start with visible swelling of the Kamenis and noticeable shaking of the islands in general. The waters of the lagoon might bubble or boil and turn white, orange or red as they are heated by the fresh magma. Those nearby may notice the pungent rotten-egg stench of hydrogen sulphide gas. After days, weeks, or maybe months of these overtures, the eruption itself would begin in earnest with a vent opening, likely somewhere on or near Nea Kameni. A lava dome will start to inflate and push out of this new conduit, perhaps yielding blocky lava flows and adding new layers to the island's architecture. This extrusion of lava will be punctuated by small explosions, sending occasional small dusty clouds, shaped like cauliflowers or twisted rope, hundreds of metres into the atmosphere – likely similar to those I saw at Santiaguito. From a distance, the hot rocks around the vent will glow like embers and will be especially impressive at night. Up close, the sound of the volcano growing will be complex, at times very loud and at other times subtle and perhaps eerie. In 1866 Fouqué described 'the incessant noise of cracks opening as the rock contracted during cooling ... accompanied by a tinkling somewhat like that of broken china.' The explosions, though, will reverberate around the islands, according to Fouqué 'like strokes of thunder mixed with shrill whistling'. Ash, from the larger explosions, might sporadically rain down on the nearby sea and land, potentially dimming or entirely cloaking the islands from the gleaming Aegean Sun that so often gilds the summer skies. If emissions are strong enough, the pungent gas cloud from the vents and fumaroles will get blown by the wind, intermittently fumigating different parts of the island and making life unpleasant at times, especially for those with asthma and other similar conditions. Vegetation and agriculture might feel the effects and the tremors could trigger larger landslides, especially down the sheer caldera walls. The delivery of supplies and transport of people onto and off the islands by air and boat will probably, at least sporadically, be disrupted.

For the two to four years that the eruption is likely to last, the whole ambience of the archipelago will be transformed from peaceful paradise to one of brooding, unpredictable power. Locals will have little choice other than to live with the intermittent inconvenience, discomfort and hazards, or move away. Whether tourists will stay away or flock to see the spectacle is hard to predict. Once it is over, we will likely need to redraft the map of the bay once again.

Studying the rocks on Santorini helps us to understand the present-day hazards here and at other systems like it. Reconstructing the largest past eruptions and their patterns at Santorini also forms part of a global database of large-scale eruption activity put together from similar studies across the world and stretching back over the millennia. As we cast back deeper into time the record becomes increasingly incomplete as the forces of weather, vegetation and tectonics obscure the readability of the volcanic debris left behind. Using these compilations of the size and timing (often determined by measuring the activity or concentrations of radioactive elements in the rocks associated with the eruption) of different types of eruption we can draw out trends about the typical recurrence rates of different sizes of event and understand the potential consequences of volcanic cataclysms of much larger scales than those recorded in our written history.

But it is not just by flinging out rocks that volcanoes impact our planet's surface and environment. Back on my first visit to the low summit of Nea Kameni, I remember noting the faint haze of gases wafting from fumaroles dotted around an old crater. Some of these are along the tourist trail, where dusty footprints trace the topographic lines left passable by the volcano's past activity and I paused to feel their damp warmth and sniff their weak musty aroma. I did not know then, but other gases seep unseen from the ground around these fumaroles. I knew that these emanations contained important information about what is going on below the surface, information that is complementary to that hidden in the rocks from past eruptions and encoded in the Earth's seismic or deformation signals. I also knew that as they mingled and blew away with the breeze, gases like these and those pumped out during large-scale events and activity had the potential to have a far wider footprint of influence on our planet than the ash clouds blown out in even big eruptions like the Minoan. Subtle as those gases were seeping out of the hot, grimy ground that day in the Aegean, I also already knew that my research would be following their trail through my PhD, studying the short range effects of such emanations on our atmosphere. As it has turned out, I have been tracing the pathways of volcanic gases one way or another ever since and I could not have imagined then how far it would take me.

Part Two

The gases of Volcanoland and their global effects

Chapter 4

Breath of the Earth

*What messages do volcanic gases
carry from the deep?*

The strange entwinement of rock and gas that drives volcanic activity like that at Masaya, Santa María or Santorini has been at the heart of theories of volcanism since their very beginnings. In fact, as we saw in Chapter 1, the force of gas rather than that of rock was often the foundation of conjectures explaining volcanic power in Western classical culture. Aristotle interpreted volcanic eruptions as the end results of the huge movements of powerful subterranean winds bursting forth. For Strabo and Seneca, winds rushed through huge underground cavities fanning the flames of Earth's underground fires. The Roman poet Ovid likened Etna to a living beast 'possessed of many air passages that breathe out flames'. I have always enjoyed, too, Kircher's very pneumatic description of volcanoes as 'the vent-holes, or breath-pipes of Nature, to give vent to the superfluous choaking fumes and smoky vapours

which fly upwards', penned in 1669. Gazing on volcanic outbursts amply illustrates the validity of the conclusions drawn by
these early authors regarding the importance of gas in volcanic
phenomena, be it Masaya's churning lava portal, flaring surges
of blazing spatter from Stromboli, throbbing fire fountains on
Etna or many other besides. Although we now find the roots of
volcanoes in the partial melting of the Earth's mantle to create
rocky liquids, without the gas held within these magmas, the
nature of our planet's volcanism would be a very different and
much more staid affair. Envisaging the terrifying Plinian eruption columns from the likes of Santa María or Santorini, their
launch propelled by the expansion of the bubbles held within
their magmas until the liquid matrix fails and breaks, brings
home volcanism's driving dynamism forged from rock and gas.
Understanding the origins and nature of the volatile chemicals
that escape from magmas as gas, frothing rock into foams and
driving huge explosions tells us both about what is going on
inside and beneath volcanoes and also, as we shall see in later
chapters, how volcanoes reach out to have effects that span
the planet.

Holding a piece of pumice in my hand at the foot of the
Bronze-Age deposits on Santorini, it was as if I could sense the
gas flowing through the rock. Experiments have shown us that
the voids – vesicles – in pumice are almost completely interconnected. It is theoretically possible to solve this miniature Minoan
labyrinth and navigate through the rock, following the steps of
the intricate dance between gas bubbles and magmatic polymers
that choreographs the different eruption styles in Volcanoland.
The primary observations are clear, and ancient instincts
regarding the basic principle still resonate with our ideas today.
But our understanding about volcanic gas has changed in many
important ways. Volcano science has pulled us away from great
flues running through the planet or internal fires belching out
soot and smoke. We know now that magma brings the gas with

it from deep within the planet, captured within its structure. At high pressures these gas molecules are held dissolved in its rocky solution. But, as the molten Earth rises and the pressure drops, these dissolved molecules are expelled and bubbles form and expand helping in turn to drop the magma's density. This drop in density enhances its buoyancy and helps to push it on towards the surface, and lower pressures that will summon even more vapour from the rocky liquid to swell its vesicles. And as we have moved beyond ideas of volcanic vapours as winds whistling through deep caverns or fire breathing beasts we have come to address other important questions too: of where these gases come from; of how they come to be dissolved in the Earth's molten rock; and of what quiet messages are they whispering to us about our planet's deep origins and processes or a volcano's shallow plumbing and restless magma.

Over my career I have measured volcanic gas emissions at many volcanic systems around the world and in many different ways. I have already described how we used a small spectrometer to measure the metallic-tasting vapours drifting out of Masaya volcano; using the subtle way the sulphur dioxide's electrons act to change the light transmitted through the plume to look up and count its molecules like stars in the heavens. I have mentioned the importance of water in terms of detonating volcanic explosions: both water dissolved in the magma and also sometimes infiltrating an eruption from external sources, like the flooding lagoonal seawaters triggering the fine, ashy surges we see recorded in Minoan deposits of Santorini. But there are many other gases belched out from the Earth's volcanic passages and vent-holes. Determining the composition of these gassy cocktails again draws on centuries of scientific endeavour. Once captured, these most ephemeral parts of a volcano's character can give us fundamental clues about what is going on within the unreachable recesses of the planet's depths.

Perhaps one of the most incongruous places that I have been on field work to measure Earth's emanations felt neither particularly volcanic nor remote, and that is what made it so unexpected. To the west of the modern-day city of Naples, not far from Pliny the Younger's vantage point in Misenum where he witnessed events at Vesuvius unfold in AD 79, is the municipality of Pozzuoli. This coastal town has a long history stretching back to Greek and Roman times. Visiting it these days, it is, on the face of it, much like many other modern Italian seaside suburbs. It has a small port and marina, the usual homes, shops, bars, metro stations and traffic. Its greater distance from the double hump of Vesuvius than the city of Naples itself might lend a traveller reassurance of increased safety from volcanic fury. But there are clues to something unusual going on peppered around Pozzuoli amid the pizzerias and apartment blocks. Great forces stir beneath the Earth here, and evidence from this bay pinching into the Italian coastline has long shaped thinking about our world and how it changes.

The land here is restless. Excavations in the 1750s near the present-day port exposed three large marble columns from a ruined Roman market building. A band of borings into these antique structures, left by marine molluscs, suggested that the site had sunk below sea level, then re-emerged. The fact that the pillars had remained upright over centuries implied this movement was gradual rather than abrupt. The explanation was hotly debated: could the land really sink and rise sufficiently gently to leave the pillars upstanding? The Scottish geologist Charles Lyell used an illustration of the columns as the frontispiece for his hugely influential 1830 tome *Principles of Geology*. His thesis was one of slow and steady geological forces acting over long periods of time and these columns were among the poster children for his point of view. A volume of *Principles of Geology* was in Darwin's travelling library on his voyage on HMS *Beagle*, with his thinking and observations influenced by Lyell's ideas of gradual long-term change with the present being the key to the past. Although much has advanced in our understanding of the world since the nineteenth century, these conceptions of big changes being possible via the action of small-scale processes over very lengthy periods, of stretching eons of geological time and of our place within it remain fundamental to this day.

Columns excavated in the Pozzuoli area (Campi Flegrei), Bay of Naples – first believed to be the 'Temple of Serapis' and now thought to be the remains of a Roman market place. Note the dark bands of borings part way up the columns showing that the shore sank below sea level, and has then been raised again. From Elogio di Francesco de' Marchi *by Francesco Tognetti (1819).*

As you look more closely around the Pozzuoli area you notice other signs of a more tumultuous geological past too. A past in which change has not, in fact, always been gradual. Cones and craters intermingle with the urban landscape. In places hot springs and fumaroles, including the bubbling mud pots and fuming crevasses of the shallow Solfatara crater, hint at something deep and powerful lurking beneath the town. In fact, Pozzuoli sits in the centre of a huge volcanic system known as Campi Flegrei (the Phlegraean Fields) – the site of Virgil's entrance to the underworld in *The Aeneid*. The bay itself is the partially flooded remnant of multiple huge caldera collapses, the most recent during an eruption, known as the Neapolitan Yellow Tuff, about 15,000 years ago. This vast eruption spread more than 50 cubic kilometres of rock over more than 1,000 square kilometres, about ten times larger than the eruption of Vesuvius that wiped Pompeii from the map. There are many parallels with other caldera-forming systems like Santorini. Like the Bronze-Age Minoan eruption, the Neapolitan Yellow Tuff event is not unique at Campi Flegrei, and there is evidence of other even larger eruptions in the deeper past. Like the historic craters, flows and domes recorded on Santorini's Kameni islands, Campi Flegrei is still very much active but currently it seems, fortunately, in a more benign phase of smaller eruptions.

Today this area is among the best-monitored volcanic precincts in the world. Scientists watch as the Earth gently swells and sinks accompanied by small earthquakes, the type of process that sedately dunked and then raised Lyell's Roman pillars. Sometimes in the past the inflating of the land has culminated in an eruption: most recently, in 1538, building the 120-metre-high cone of Monte Nuovo in just one week. Uplift in the 1500s was gauged by the noticeable rise of land metres out of the sea. We have more precise measurements now. In the 1980s a rise of almost two metres, attended by a shallow earthquake swarm, led to the evacuation of 30,000 people due to the perceived risk of imminent eruption. The unrest calmed without further incident but this restless giant continues to keep us on our toes.

I first journeyed to Pozzuoli in the summer of 2002. Our mission was gas sampling and I have to admit that, despite Vesuvius's looming presence 25 kilometres to the east, my initial impression of the place was not all that promising in terms of measurements of volcanic gases. As low as the altitude of Masaya is, it at least had greeted me the December before with a visibly puffing summit. It felt odd in contrast to be driving our sampling equipment through the everyday bustle of this seemingly normal Italian town. We parked our small hire car on the street in front of an ornate pink three-storey building which, attractive at it was, spoke nothing to me of volcanism. Nonetheless, a yellow sign on the grey wall at its base advertised VULCANO SOLFATARA ENTRATA with a red arrow pointing helpfully to an archway under the left panel of the building's balconied triptych front. I remained silently sceptical but pulled my pack from the car and trotted along with the others through the archway's portal towards its alleged promise of volcanism.

My incredulity was soon proved wrong. Although the track took us through low trees dotted with the tents and caravans of a regular-seeming tourist campsite, a strange rotten-egg smell invaded my nostrils. The scent was sufficiently pungent that I was not sure that it would be my first-choice location to pitch my tent. Shortly after, we emerged from the scrubby trees into a pale moonscape hidden and quite unexpected amid this bustling town. The shallow irregular crater of Solfatara is just a few hundred metres in diameter, a pallid scar of dusty volcanic rock against the green of the surrounding gardens and scrub land, kept clear of vegetation by the seeping fluids from below. In some places, mud pots bubble and gassy clouds waft gently from the Earth. In others, they hiss out at high pressure and boiling temperatures from fumaroles like the Bocca Grande. The chemistry of these fluids is complex and hosts unusual extremophile microbes feeding off the sulphur despite the high temperatures and acidities. The rocks around the fumarole mouths are colourful: vivid orange vents, grading to a paler yellow, dusted coronas as the steeply cooling temperatures shift the subtle precipitation chemistry of mineral dust settling from the fumes. A palette of colours created by the interweaving of sulphur, with elements like potassium, aluminium, iron and arsenic. In the vaulted cavities of the ruined bath houses adorning the eastern crater wall, needled crystals of sulphur compounds grow like snowflakes into tiny brittle structures lacing the brickwork. But despite this unexpected and delicate beauty, danger lurks here: in 2017 three people died falling into a boiling cleft in the crater floor, and the earthquake swarms associated with Campi Flegrei's unrest are often centred around or directly under this crater.

Back in 2002, we were there to meet one of the scientists from the local monitoring agency to help with their ongoing programme of measuring the gases. Having set my own equipment up, I watched as he hammered a metal tube into the angry orange mouth of the Bocca Grande, attaching a series of taps, glass tubes, bulbs and flasks and skilfully using the fumarole's own pressure to draw the gas through these different receptacles for lab analyses. It was fascinating to watch this intricate work with the low hiss of the fumarole in the background amid the strange dissonance of the landscape. If I had studied the history of my new-found discipline in more detail before that early trip, I would have realised that those measurements that hazy July day were part of a long history of gas measurements in and around the Bay of Naples. Thanks again to the long history of recorded human thought in this area of the Mediterranean, we know more about observations and ideas concerning volcanic gases here than we do in most, probably all, other parts of the world.

Early gas measurement techniques were far more physiological than our current chemistry and spectroscopy, with vapours often characterised by their effects on animals or humans. In AD 77, Pliny the Elder, the same thinker who later perished in the AD 79 eruption of Vesuvius, records the presence of 'Charon's* passages, holes releasing deadly vapours' in the Pozzuoli area causing death 'to any animals with the exception of man'. There is ambiguity regarding the exact location of Pliny's vaporous cracks, but later interest seems to have focused around, to use the words of the Italian writer Benedetto di Falco in 1535, a 'small and miraculous cave'. This cave reeked 'with a strong smell of sulphur or another unknown quality of the soil, so that any animal entering the cave suddenly dies.' By the 1600s the cave had become known as the Grotta del Cane (Cave of the Dog) after a local party trick (and tip-earner) whereby a small dog would be held in the cave until unconscious and seemingly dead and then revived by being thrown into the nearby Lake Agnano. Athanasius Kircher, whose vision of Earth's internal furnaces from his *Mundus subterraneus* was described in Chapter 1 and whose account of the birth of Mikra Kameni from Santorini's lagoon was chronicled in Chapter 3, first visited the cave in 1638 and observed the trademark anaesthetising and revival of the dog. Kircher was intrigued by the cave's effects. Wanting to know more he put his own head into the cave and immediately felt the noxious gas take hold; he swiftly pulled himself out before he succumbed.

The cave obviously made an impression on Kircher and he returned decades later in 1664 with a contingent of more than 50 illustrious witnesses, including doctors, counts, princes and bishops. This eminent party proceeded to make and record systematic observations of the grotto. They repeated the dog experiment, noting that once the dog was revived 'it rose, looked

---

* The ferryman of Hades, who was believed to carry souls of the newly deceased across the river Styx, which divided the world of the living from the world of the dead.

around and ran away to avoid a repeat performance.' Against the advice of the others, one doctor of law entered the cave with a lighted torch and established that the flame went out when lowered below a certain level and reignited when lifted again. Through this eclectic mixture of expertise and investigation, they established that the invisible smothering fumes were heavier than air but lighter than water. The Marquis de Arenis intended to obtain a sample of the vapours, but his servants forgot to bring a container ('as happens everywhere', according to Kircher). I wish I had others to blame when I forget vital pieces of field kit.

Knowledge of different gas chemistries and analytical techniques advanced over the following decades, and in the late 1700s and early 1800s the cave's mysterious nature yielded to science. Around 1798, the Abbé Lazarro Spallanzani, Professor-Royal of Natural History in the University of Pavia, collaborated with Scipione Breislak, a priest teacher at the Military Academy of Naples, and established that the noxious vapours were ten per cent 'vital air or oxygenous gas' (oxygen), 40 per cent 'fixed air or carbonic acid' (carbon dioxide) and 50 per cent 'phlogisticated air, or azotic gas' (nitrogen). The carbon dioxide seeping up through cracks from the Campi Flegrei volcanic system, being denser than air, settles onto the cave floor like an invisible tide rising from below. This asphyxiating layer snuffs out flames and animals if they are left in it too long.

Although gas-analysis techniques had moved to something more humane than poisoning helpless creatures, it seems that the dog asphyxiation stunt persisted as a trick for tourists. Mark Twain was especially, and rather macabrely, excited to participate in this particular chemistry experiment. In his 1869 travelogue *The Innocents Abroad*, he wrote of his longing to fully suffocate a dog in the cave himself. But upon arriving at the grotto, in his words, 'an important difficulty presented itself. We had no dog'. Perhaps his report of the dog's death would have, in any case, been an exaggeration. Happily for the local dog population, shortly before the Second World War the authorities banned such abuse. Less happily, the cave fell into neglect and disrepair until recently, when rumour has it visiting is again possible thanks to the efforts of local volunteers. Presumably, however, it's no dogs allowed.

Volcanic activity around the Bay of Naples continued to draw the attention of this nascent field of gas geochemistry. In late 1834, Charles Daubeny, professor of chemistry and botany at the University of Oxford, visited Vesuvius after a particularly large eruption of lava buried 180 houses and 500 acres of land, displaced 800 people and swept away one wall of the Casino of the Prince of Ottajano. Daubeny recounted his experiences the year after to the Royal Society in London. The eruption was quiet by the time he arrived, but he relays the testimony of a local geologist, Teodoro Monticelli. In Monticelli's accounts the volcano throws up stones and scoria; 'smoke' billows; the ground quakes; conical hillocks rise, break and sink; currents of lava bend down the mountain, sometimes pouring down guided by 'the boundaries of a hollow way or water-course' or a road, following the steepest plumb lines in the topography; clouds of black sand hover over the flowing lava punctuated by 'frequent coruscations of lightning'.

The volcano's 'state of comparative repose' devoid of 'signs of internal commotion' allowed Daubeny to investigate the mountain's activity and to collect 'solid as well as aeriform substances'. While some deeper parts of the new crater were 'so charged with the noxious vapours evolved from an infinity of minute and scarcely visible spiracles, that it was judged unsafe to venture down', Daubeny and his guide could safely access the shallow parts to collect flaky sublimates and condense the vapours still bursting forth from its crevices. Daubeny traced the strata of ash and lapilli around the crater noting their saline coating of 'common salt coloured red and yellow by peroxide of iron' rendering their appearance 'beautiful and brilliant'. To collect the gas, he had built his own field apparatus consisting of a metal tube that could be pushed a little way into the ground and fed into the head of a large alembic — an alchemical still — to capture the vapour for analysis. The crater gases contained large quantities of steam, muriatic acid (hydrochloric acid) and carbon dioxide. Similar measurements down on the lava flows yielded steam, hydrochloric acid and ammonium chloride.

Fortunately for the present-day Neapolitans, Vesuvius has not thus far been active within my lifetime and I have not yet had cause to follow in Daubeny's footsteps from Oxford's dreaming spires to make gas measurements on this remarkable mountain. But he also visited Solfatara and, like me, took samples from the hissing fumaroles and bubbling mud pots. In fact, his alchemical apparatus does not sound dissimilar to the metal and glass set-up I watched in use in 2002. In his address to the Royal Society he made systematic comparison between the emanations from the 'semi-extinct volcano' of Campi Flegrei, the fresh activity on Vesuvius and what was then known about other ancient volcanic products. Some of his explanations fall wide of the mark, but reading again his 1835 paper I am struck by the way he drew key observations that remain relevant to our understanding of volcanic degassing today: the prevalence of hydrochloric acid emission from slowly cooling lava flows; the strong emissions of hydrogen sulphide (the rotten-egg smell that I had noted) from Solfatara and its absence from the emissions at Vesuvius; the presence of sulphur compounds only in the products from the earlier stages of Vesuvius's activity; the comparison of his measurements of actual volcanic gas emanations with the water and hydrochloric acid known to be held in the rocks from ancient eruptions like those in the Auvergne; and the infiltration of volcanic gas streams by atmospheric air and water that 'find their way to the seat of volcanic operations'.

Ferdinand Fouqué also knew the value of measuring volcanic

Solfatara crater with the city of Pozzuoli and Monte Nuovo (erupted in 1538) in background

gases. He recorded their chemistry at Vesuvius and the Aeolian Islands for his doctoral thesis and measured these emanations in 1866 on the Kameni islands. His book on Santorini gives chemical recipes for detecting his different volcanic target gases. He describes how to sample gases from fumaroles on land and those bubbling up through the waters lapping the shores using tubes, funnels, valves and glassware. Having read his careful descriptions, I did pause to wonder what Fouqué's verdict would have been about our improvised collection of the gas fizzing up in Nea Kameni's shallows using an inverted children's paddling pool purchased at a Fira toyshop during the 2011 unrest of the islands.

But this sampling from cracks in the Earth, waning lava flows and bubbling waters left many questions about the nature of volcanic gas. At the turn of the twentieth century, although many of the components of volcanic gases seemed clear – carbon dioxide, the sulphurous gases of sulphur dioxide and hydrogen sulphide and chlorides like biting acidic hydrogen chloride – there was still some debate about whether water played a significant role in volcanism. This is perhaps surprising given the billowing white steam-like plumes from volcanoes and the often-invoked explanation of volcanic eruptions as massive steam explosions. As you may remember from Chapter 2, Alfred Lacroix developed an analogy of magma's dissolved water vapour driving eruptions like champagne from a shaken bottle while gazing on the fast-moving clouds bursting forth during the smaller-scale activity that followed the terrible eruption of Pelée in 1902. Lacroix was also, as it happens, Fouqué's son-in-law. The patriarch's dedication to science can perhaps be gauged by the rumour that he refused Lacroix permission to marry his daughter until he had completed his doctoral thesis.

But despite these persuasive observations supporting the importance of steam in volcanic gas plumes, there continued to be influential dissenters. The Swiss chemist Albert Brun studied volcanoes from around the world and in his 1911 book *Recherches sur l'Exhalaison Volcanique* (*Volcanic Exhalation Research*) dismissed the presence of water in no uncertain terms. For him the large white plume (he, of course, uses the more stylish French word '*panache*') – characteristic of volcanic activity – was clearly composed of solid particles of salts. Brun believed unequivocally that the aqueous theory should disappear from science. Measurements of gases emitted directly from fresh lava itself were key to addressing this controversy and the story of the first such measurements takes us very far away from the shadow of Vesuvius or indeed the terrible fury of 1902 Martinique, to a fragment of volcanic land, remote, in fact, from most places, namely Hawaii. Distant as Hawaii is from Italy, there is, as we shall see, a direct line of volcanological heritage stretching from the Bay of Naples via the Caribbean to the rim of Kīlauea, one of this Pacific archipelago's most active volcanoes. A line of heritage that we, too, tread now and a journey that will reveal more about what gases come out of volcanoes, where they come from and what they tell us about our planet's inner nature and ancient history.

There are many special things about the islands of Hawaii. When I visited in 2008, we did not have as much time off to explore as I would have wished for, but what I do remember about our free time seems to mainly revolve around turtles. We spent one incredible afternoon on Punalu'u beach mesmerised by the population of green sea turtles. We watched from a respectful distance as they laboured up the palm-fringed beach and buried themselves in the black sand, flicking it back over themselves with their fore flippers. Where the waves lapped the shore, their domed shells mingled with the dark volcanic boulders as they wallowed in the shallows. On my final day I even got to snorkel following a turtle as it darted effortlessly amid the vivid colours of the coral, our roles now reversed, with me feeling as lacking in grace in the water as the turtles had seemed laboured on land. Unfortunately, despite the copious application of sun cream, I paid for this pleasure by badly burning the backs of my legs and arms in the tropical sun, making the lengthy flights back to London very much more uncomfortable. It was nonetheless entirely worth it.

Hawaii is a very special place volcanically. This long line of islands sits in splendid isolation in the middle of the Pacific Ocean. The nearest continents are thousands of kilometres away. If we were able to peer beneath the waves there is no deep trench or ridged rupture telling of two tectonic plates meeting and subducting or moving apart. Instead the great volcanoes of Hawaii rise majestically from the sea floor right in the heart of the Earth's largest tectonic plate – the Pacific. The islands' impressive shield-like mountains are but the tips of something far more impressive still in terms of magma's ability to build as well as destroy. Mauna Loa (Long Mountain) on Big Island Hawaii is one of the most voluminous volcanoes on Earth, rising to over four kilometres in elevation above the surf and with its gently sloping submarine flanks sinking an additional five kilometres to the sea floor. Mauna Loa's mass is so great that the Pacific plate flexes under its weight, sagging like a hammock beneath the island.

Mauna Loa is one of five volcanoes that make up Big Island and, apart from its much smaller south-easterly neighbour Kīlauea, it is the island's most active. Tracing the line of the Hawaiian islands away from Big Island to the north-west, the volcanoes become lower and older, with longer and longer periods since their last eruptions. The land of the Midway and Kure atolls, over 2,000 kilometres from Big Island, rises to just a few metres above the waves – rings of shelly sand and coral peeping out of the Pacific, home to sea birds and strategic airstrips. This chain of land – a tactical staging point between the Americas and Asia – is built by the third main type of volcanism on our planet – hotspot volcanism. Unlike the great arcs and ridges of volcanoes that outline where plates split apart and meet, under Hawaii a plume of hot mantle rises from deep and partially melts as the pressure drops, making magma. Some of these magmas find their way up from the mantle and through the thin ocean crust of the Pacific, erupting mainly as runny low-viscosity basalts.

This 'hotspot' in the mantle remains stationary while the rigid plate moves above it. Volcanoes gradually push above the waves over millennia, fuelled by the hotspot's magmas while they sit above it. But time passes and the Pacific plate pushes slowly north-east and carries the once active volcanoes away from the hot area of underlying mantle. Without the hotspot feeding them with magma their activity ceases, leaving a stream of extinct volcanoes trailing from Big Island to Kure, and beyond, tracing the path of the plate's movement. This

explanation for the formation of the Hawaiian chain was proposed by Canadian geophysicist J Tuzo Wilson in the 1960s
to support the unifying revolution of plate tectonic thinking in
geosciences. Hawaii is not unique; these mantle hotspots are
scattered across the planet. Some, like those under Hawaii and
the Galápagos islands off South America, are in the middle of
ocean plates. Others, like Iceland in the north Atlantic, coincide
with tectonic plate boundaries raising land from the sea where
the Earth is splitting. And there are hotspots as well under the
continents, like that feeding Yellowstone, or implicated in splitting continents apart like the one under east Africa (that we will
discuss more in Chapter 6).

In the Pacific, once the sustaining supply of magma is cut off
to an island and the rock beneath it cools, the processes of erosion via wind, rain and waves, collapse and gradual sinking will
pull the land back into the ocean. Slowly the islands diminish
until the rings of their coral reefs are their main feature forming
an atoll above the waves. With time the rocks recede deeper still
into retirement as the bump of a seamount on the ocean floor.
What we see now is just another snapshot in geological time.
Rocks from along the train of islands, atolls and seamounts
can be dated by measuring the argon gas left trapped in them
from the slow radioactive decay of the rock-forming element
potassium. By this measure the ancient rocks making Midway
Atoll, just over 2,400 kilometres from Kīlauea, were erupted
approximately 28 million years ago. This simple speedometer
clocks the slow drift of the Pacific plate at about 8.5 centimetres
per year north-west over the mantle's hotspot towards its ultimate death in the subduction zones of the Kuriles, Japan and
the Marianas. One day Mauna Loa and Kīlauea will drift off
beyond the hotspot's reach and fall silent too. Unimaginable as
it seems today, one day they will sink back beneath the waves.
In another 28 million years, depending on the longevity of
the Pacific plate's current trend of motion and the Hawaiian

hotspot, we might imagine Big Island as an atoll like Midway is today. To the south-east new mountains will have risen from the sea floor above the waves. This process has already started with the chain's newest volcano, Kamaʻehuakanaloa (formerly Lōʻihi Seamount), yet to break above the sea off Big Island's south-eastern shore. Again the Earth's magma forms a lifeblood helping to drive our planet's constant cycles of ebb and flow, destruction and renewal.

But Hawaii's volcanological significance is not just as a prime example of a hotspot volcanic chain. Thanks to the almost constant and relatively safe volcanic activity of Mauna Loa and Kīlauea over recorded times and to the important work of the Hawaiian Volcano Observatory (HVO), this remote part of the world has become an important field laboratory for volcanologists. This has included the sustained and systematic study of volcanic gas emissions.

The founding of HVO in 1912 is credited largely to the vision and passion of Thomas Jaggar. Jaggar was possessed by volcanoes and for good reason. At the age of 31, while an instructor in geology at Harvard University, he was one of a small band of scientists sent from the US to investigate the 1902 eruption of Pelée. Here he overlapped with Tempest Anderson and Alfred Lacroix, sharing in their horror and awe at the power of the volcanic forces at work and the challenges of giving sound advice to the officials charged with managing the danger to the local population. He was reportedly the first to spot the strange monolithic 'spine of Pelée' as a sticky viscous lava plug squeezed its way out after the eruption to preside over its own devastation.

The experience understandably left a deep mark on Jaggar. He wrote of the dreadfulness of death everywhere, of finding a dead baby in a cradle and of a baker cooked inside his own oven, where he had taken refuge from the eruption. He was haunted by the odour that returned in dreams, 'a combination of foundry and steam and sulphur matches and burnt things, every now and then a whiff of roast, decayed flesh that is horrible'. He interviewed some of the few survivors and years later reflected that these human contacts had made the greatest impression. Gradually he 'realized that the killing of thousands of persons by subterranean machinery totally unknown to geologists and then unexplainable was worthy of a life work.'

In 1906 Jaggar travelled to Vesuvius hoping to witness the eruption that had caught his eye in a newspaper headline. Here he met, 'picturesquely plastered' by dust and ash, fellow American Frank Perret and Raffaele Matteucci, the director of the Vesuvius Observatory. An engineer by training, Perret's passion for volcanoes had been sown during his childhood by a fascination with an engraving of the destruction of Pompeii in his family store in Brooklyn and by the vivid colours of the setting sun in 1883 attributed, as we will discuss in Chapter 5, to the Krakatau eruption in Indonesia. Following a breakdown in his health, he moved to be close to Vesuvius and became Honorary Assistant to the Royal Observatory where he joined in the work of watching the volcano. This included, as described earlier, setting his teeth against his iron bedframe to feel the volcanic tremor.

Matteucci, too, was devoted to Vesuvius. He was once apparently quoted as saying of the mountain that 'she and I dwell together in solitude mysterious and terrible.' In 1900 the affair had nearly ended badly when he was hospitalised after a minor explosion rained hot rocks on him at the summit crater. He had apparently started to run and then turned to watch the action. This story reminds me of my first ascent of Stromboli volcano, in 2002. Just before we reached the closest point of our path to the active vents, my more experienced colleague turned to me and offered the advice that if a more violent explosion did throw rocks our way I could either turn my back and run like crazy or face them and try to dodge the larger chunks. 'Which one would you go for?' I enquired. 'I am generally in the run like crazy camp,' he answered. Fortunately, I have not so far faced this dilemma and I hope never to do so.

Although Jaggar technically missed the actual eruption in 1906, his Vesuvius visit, the observatory and Perret's various innovations in terms of monitoring instruments nonetheless impressed him deeply. His obsession was further fuelled in 1907 when he travelled to Alaska and observed dramatic changes at the volcanic island of Bogoslof. Part of the island had risen by almost 8 metres in less than a year. He became convinced that to answer the fundamental questions about how these great forces were unfolding and to save people from catastrophes like that on Martinique, a global network of geonomical observatories was needed to study the Earth's internal activity in much the way that astronomical observatories exist to chart the heavens. Despite his global ambitions, in 1909 he was to find his specific volcanic muse. On the way to Japan with his wife, Helen, they took a detour to see Hawaii's famous lava lake at Kīlauea. This pool of molten rock has been present on and off at the volcano's summit since at least its first descriptions by missionaries in 1823 and certainly far deeper in time than even Hawaiian legends delve: 'from chaos till now' (in the words recorded from the testimony of the native Hawaiians by William Ellis, an early missionary).

In 1909 the journey from Hilo to Kīlauea's summit involved a day of travel, by steam train and stagecoach through fields of sugar cane, moss-covered forests festooned with hanging vines, native 'ōhi'a trees and giant ferns and finally out into the open nearer the volcano's summit. Steaming cracks welcomed the Jaggars to Halema'uma'u crater. Jaggar trekked to visit the lava lake every day of their three-day sojourn at the summit, joined by Helen on the final occasion. Along the rough trail of sharp congealed lava across Kīlauea's caldera floor, visitors could pause to singe postcards in hot clefts as souvenirs. The lake itself was within the inner Halema'uma'u crater and, if more recent activity is a guide, was probably oval in shape and something like 100 metres across. The roiling energy of the lake impressed Helen: 'There isn't a second's pause in the ebb and flow of the lava, the satiny, black-grey crust forms only to be broken at once by the lightning-like cracks in the surface exposing the most livid-flame coloured molten lava.' The colour scheme of the volcano's gases seems to have also captured her imagination: 'All around the edge of the lava lake rose sheets of steam, tinted, by the colour of the lava, to the most lovely shades of pink from delicate shell pink to deep rose.' For her husband this first encounter with the Halemaʻumaʻu lava lake was a profound experience. In Kīlauea he had found a seemingly perpetually active volcano, with magma on display at the surface and relatively safe to approach at all times of the year. In Honolulu he met with members of the Hawaiian business community willing to champion and fund a permanent monitoring station. During his onward travels in Japan he encountered Fusakichi Omori, who had invented a new scientific instrument to record earthquakes and ground tilt. Jaggar would write of this trip to the Pacific that it was 'as if everything within me converged.'

And so began Jaggar's scientific work on Hawaii's volcanoes. Initially he was not always able to be at Kīlauea's side, but he had help from old friends. In 1911 Perret came to stay and participated in an intricate experiment to measure the temperature of the lava in the lake that became headline news in the Pacific Commercial Advertiser ('Temperature of lava recorded – heat 1010 centigrade'). Such heroics made Perret something of a celebrity. For several months he set himself up living in a wooden hut at the crater's edge in order to keep the lava lake under constant surveillance. Visitors would come and seek him there: a 'strange little man who seemed to thrive on air filled with the smell of sulfur' according to one. Having sat for many hours in volcanic plumes, including that of Halemaʻumaʻu, I can confidently say that my dedication to volcanology does not extend to setting up house in such an environment. For Jaggar, there were still many snags to negotiate, but finally, in 1912, he moved to Hawaii and Kīlauea became his laboratory. Shortly after, his marriage to Helen ended and she returned with their children to the mainland US.

The first gas measurements were made on Kīlauea in 1912. Jaggar was away from the islands at the time attending to a family emergency in Boston, and the measurements were led by Arthur Day and Ernest Shepherd from the Geophysical Laboratory of the Carnegie Institute of Washington. Their study was a direct response to Brun and his dismissal of water as a key volcanic gas. Their aim was to collect gas fresh from the liquid lava free of contamination from atmospheric air, an endeavour deemed more likely to be feasible at Kīlauea than any other known volcano on the planet.

They made the 'somewhat difficult descent into the crater without mishap' on rope ladders. The gases must have been choking and they wore rubber nose masks stuffed with a water-soaked sponge to absorb them. With some difficulty they pushed an iron pipe into the domed carapace over a spattering column of liquid lava that had worked its way up through the shattered crater floor and pumped the gases through a series of glass tubes. Water 'clouded with free sulphur' began to condense in the tubes 'with the first stroke of the pump'. Glassware is poorly adapted for volcanic sampling and only eight out of their 20 tubes made it back intact without cracks. Having had similar frustrations during volcanic gas sampling I feel their pain. Nonetheless the campaign had been a success and they had collected 'a quantity of water sufficient to establish its existence among the volatile ingredients exhaled by the volcano beyond the criticism of the most sceptical.' They attempted further measurements some months later with the descent made at night-time so that they could spot the volcanic emissions more easily via the 'pale blue flame of the escaping gases', likely burning sulphur and hydrogen, 'plainly seen emerging from the crack in the dome.' Although less successful, these repeat nocturnal measurements were additionally heroic given that two days after their first incursion into the crater, part of its floor had collapsed and their sampling station had sunk into the fuming abyss.

Once established on Hawaii, Jaggar also made gas measurements using similar methods, including importantly for some of the more recent work that I shall discuss later, the 1917 summit lava lake. These early measurements extend our charting of the chemistry of the hot exhalations from Kīlauea over a far longer period than most volcanoes globally. Although the gas mixtures measured have subsequently needed correcting for contamination by the Earth's oxygen-rich atmosphere, we still refer to this data today. These days we have many other methods of sampling these high-temperature vapours – many of which, fortunately, do not involve brittle glassware. Acidic gases like sulphur dioxide and hydrochloric acid can be collected by pumping volcanic fumes through jars of watery alkali liquids or filter papers soaked in these alkali solutions. The chemistry (acid plus base goes to salt and water) is drilled into many of us at school and I often return from the field with great bundles of labelled Ziploc bags with these tissue discs ready for analysis. When I returned from Nicaragua in 2001, I went straight home for family Christmas and had to beg room to store them in my parents' freezer to help to keep them fresh for analysis. This took valuable space from the turkey leftovers but my mum did enjoy telling people that we had volcanic gases in the freezer. The visitor who asked to see them could not help but admit that they were rather underwhelmed. I think that at the very least they were expecting something more colourful than white discs of filter paper.

To study volcanic gas these days we do not always need to enter the volcanic plume at all. Since the second half of the twentieth century, we have developed field spectrometers, like the one that I was using on Masaya in 2001. These can use the way that volcanic emissions alter ultraviolet, visible or infrared radiation passing through them to count the different molecules present. Sometimes, like at Halema'uma'u's lava lake in the 1960s, it is possible to use the infrared or heat radiation from the molten Earth itself to look at these gas compositions right where they are gasped from the magma. Since the 1970s, thanks initially to satellites designed to map stratospheric ozone concentrations, we have been able to observe, from orbit, higher-altitude injections of volcanic sulphur dioxide, although routine satellite measurements of other types of volcanic gas remain more elusive. Sometimes, more recently still, volcanologists have mounted gas monitoring equipment on airborne drones allowing them to reach gas emissions that were previously inaccessible due to steep, unstable volcanic slopes or dangerous activity.

This accelerating effort has yielded many insights. We know the basic menu of subterranean ingredients from which a volcanic plume lofting from a summit or hissing fumarole will draw. We expect a good portion of steam and carbon dioxide and these gases generally dominate the mixture. As I have described, we can often smell the stinking sulphur gases: acidic sulphur dioxide smelling a little like burnt matches and the rotten-egg reek of hydrogen sulphide gas. As at Solfatara's camp site, this smell can be particularly pervasive in volcanic areas. On the island of Vulcano in Italy you can smell it in the water when you shower. On Vulcano, it is only a short walk from the beach up to the volcanic crater and I have watched tourists in flip-flops and sarongs turn back in disgust at the first whiff as I worked protected by my gas mask. Some years ago, after I gave a particularly vivid account of the different smells of volcanic sulphur emissions during a radio interview, a journalist asked if I considered myself a connoisseur of these gases. I had never thought of it this way, but I liked the image of me sampling

volcanic aromas with great ceremony before offering my considered verdict. It would also be a wonderful development if I
could calibrate my nostrils to return quantitative information to
save my back bearing the sampling equipment up these fuming
mountains. Sadly, this is almost certainly beyond our human
powers as well as highly incompatible with health and safety.

Beyond water, carbon dioxide and the sulphur gases, there are
other more minor ingredients in the pungent chemical cocktail
of volcanic plumes. Take, for example, the acidic halogen gases:
hydrogen chloride, fluoride, bromide and iodide, that we can
ever more routinely measure despite their low levels. Hydrogen
fluoride is especially aggressive and once etched fine lines into
a colleague's glasses during a particularly intense day at the
summit of Etna. These gases cling to your hair and the fibres
of your clothes. There are times that I have felt so imbued by
acidity after a day of sampling that I have gone straight into the
shower fully dressed to try to soak it out of me. Cotton clothing
can be slowly eaten by the grasp of these mordant molecules.
Many of my T-shirts have developed alarming holes in the wash
after exposure. A colleague has a tale of the seat of his trousers
disintegrating on the way home after a day's sampling sat among
the fuming cracks in the ground.

There are many other chemicals that find their way out of
the Earth in volcanic plumes. I sometimes ponder that in one
form or another volcanoes spread almost the full roll call of the
periodic table onto the planet's surface and into its atmosphere.
Although not all these elements are concentrated in volcanic
gas, there are many messages hidden within these emanations.
Compiling gas data from volcanoes around the world allows
us to spot patterns. Some of these messages are about things
that changed the vapours at relatively shallow levels below their
volcanic vents. We can use gases like nitrogen and argon to
spot contamination by the air or shallow water bodies drawn
from rain, snow or ice. We can see the extra carbon released

from volcanoes like those in the Mediterranean as their rising magmas heat and decompose the limestone foundations upon which they sit.

But much of the gas is from much deeper than this, imprisoned in the magmas from their birth within the Earth's mantle. Sometimes the gases carry messages from deeper still. In the case of hotspot volcanoes like Hawaii, where the plumes of hot material balloon from deep within the mantle (perhaps even as deep as the core-mantle boundary), the balance of isotopes – atoms of the same element with different masses – in the helium gas seeping out speak of tapping primordial reservoirs locked away and preserved over the eons. Such signals wafting from the ancient remnants of Earth's earlier composition remind us that solar system scale processes playing out many billions of years ago drew the blueprints of our present-day planet, including the balance of the volatile gases that bubble from the magmas and creep or hiss out through fumaroles at volcanoes like Kīlauea, Campi Flegrei and Etna for me and others to sample and deliberate over as we try to make sense of Volcanoland.

There are many mysteries yet to be solved about how our world came to have the composition that it does. Geochemists try to learn what they can by comparing our planet's bulk chemistry, and that of its constituent parts, to other celestial bodies like the Moon, the Sun, comets and meteorites. Like the other planets of our solar system, Earth formed from a disc of gas and dust rotating around our young Sun. Proto-Earth's initial composition was set, in part, by its location in the hotter portion on this turntable of spinning debris, relatively close to the growing young star. The warmth of Earth's proto-galactic environment sets limits on the quantities of volatile substances like water – with low melting and boiling points – that we think would have been available to our nascent world, as they need to be condensed as liquids or solids to be retained. Based on what we know about the chemical affinities, in other words the other elements that these volatile elements readily bond with, some other early processes like the separation of our metal-rich core from the rocky part of the Earth, also likely locked away some volatiles in appreciable proportions. We cannot probe the core's composition directly and so it is hard to know for sure, but significant quantities of prevalent ingredients in volcanic emanations like carbon and sulphur are likely sequestered in these deepest planetary shells. Larger collisions with other planet-sized bodies, like the huge impact proposed to have formed our Moon, can lead to volatile loss or gain depending on what we assume about the exact circumstances. In fact, many of our current models conclude that a significant fraction of Earth's volatiles were not endemic to this protoplanet at all. Some studies suggest that the majority of the steam molecules that seep, bubble and explode from today's volcanoes were delivered piecemeal from the colder outer solar system via meteorites or comets crashing into the planet perhaps as late as 200 million years into our world's early history. There is much still to work out.

Water is one of the most characteristic substances of our blue planet. About 70 per cent of Earth's surface is covered in oceans and it is fundamental to all the life that it hosts, including us. It is sometimes sobering to contemplate that the plan for our water world was perhaps not somehow irrevocably written into Earth's earliest conception and that we still do not fully grasp what processes bestowed its bountiful abundance today. Further, creating a surface environment suitable for life like us to thrive is not just about how much water Earth got bequeathed. How it has been moved between the different parts of the planet since is of vital importance also, and volcanoes play a crucial part in mediating some of these slow aqueous transfers between the inner earth and its surface. Part of working out how water cycles on and through the Earth involves knowing how much is in its different reservoirs. We can look around us to estimate how much is in surface reservoirs like the oceans, atmosphere, soils and ice caps.

Working out how water moves between these stores is the job of hydrologists, meteorologists and oceanographers, and shifts in how it is balanced between them can have acute consequences on the human scale in the form of floods, droughts or violent storms. But Earth has another deeper, longer-term water cycle, too, which underpins our surface hydrology and involves the Earth's inner workings as well.

It is hard to know how much water lurks in Earth's inaccessible mantle. We know that there is some held in its mineral depths and that variations in its concentration might affect the mantle's viscosity, but what fraction or multiple it is of that found on the surface remains mysterious. Whatever percentage of an ocean is contained in the fathoms of rock beneath our feet, we do know that some of it is continuously leaking out of the world's volcanoes. At mid-ocean ridges and hotspot volcanoes, where magmatism is driven by hot mantle rising to lower pressures and melting, it follows that most of the water dissolved in their magmas will be derived from the planet's deeper abyss. Given the eons of Earth's existence this might beg the question of why this constant outgassing hasn't seeped away the water from at least the shallow mantle, potentially flooding our world, heating the surface via a steam-driven greenhouse atmosphere and desiccating the planet's innards. The fact that this is not the case hints at some balance in the grand volatile cycles of our world. This is lucky for us, as if it was not so it is less likely that we would be here to try to understand these overarching processes. We might not routinely think of the deep storeys of the solid planet beneath us being a part of our lived environment, but they are fundamental to forming and maintaining our world today, even if we do not directly experience their actions in the way we do the power of wind, waves or weather.

As well as being outgassed from volcanoes, water and other volatile elements like carbon and sulphur are pulled back into the Earth's interior too. This largely happens at subduction zones where, as I alluded to in Chapter 1, these chemicals are tied to the sinking tectonic plates in multiple different ways: in the sediments that form on the ocean floor; bound up in the structure of the new crust at its very creation at mid-ocean ridges; or even locked into the plate's rigid layer of mantle by infiltrating fluids. We see evidence for this process in the compendium of gas compositions measured since the days of Daubeny and Jaggar. The gases from the volcanoes born at subduction zones are generally enriched in water and chlorine compared to their mid-ocean ridge and hotspot counterparts. These chemical elements are drawn into the Earth from ancient seawater, riding in and on the sinking tectonic plate. While some proportion of them will get dragged down into the planet's depths, it is, as we discussed before, the release of some fraction of these substances, especially water, into the overlying mantle wedge that sullies the pristine mantle (like adding salt to ice) causing it to melt. These volatile elements become incorporated into the very magmas that they spawn, only to find their way out again as they rise to the surface and are exhaled, escaping through fumaroles, lava lakes or other volcanic degassing phenomena and adding to the potential explosivity of subduction zone volcanoes.

It is certainly possible to imagine scenarios whereby a planet's outgassing and ingress of water are not balanced. Water could collect at the surface, flooding the land and the atmosphere. Or water could accumulate in the planet's innards and dry out the oceans. The relatively constant total inventory of Earth's surface hydrosphere shows that, luckily for us as water-dependent but land-based lifeforms, this planetary-scale aqueous outgassing and intake via processes like volcanism and subduction has remained balanced for much of our world's history. We will discuss the importance of these planet-scale volatile cycles more in Chapters 6 and 7. There are certainly other ways that we can imagine a planet maintaining some sort of long-term balance without Earth's particular brand of plate tectonics, and there is still much to work out about these enigmatic processes. Fitting volcanic outgassing and subduction into Earth's specific framework of long-term life support, however, might lead us to contemplate that the slow creep of the solid Earth's convection deep beneath our feet might not be a chance circumstance that coincides with our existence, but may, in fact, be a pre-requisite.

Contemplating a bubbling lava lake or watching booms of fiery explosions can certainly leave you with a resounding sense of something primeval and profound about the nature of our planet. But as well as telling us fundamental things about our world, the messages held in volcanic gases can also give us important information about the immediate subsurface plumbing of a particular volcano and the hazards that it presents. My trip to Kīlauea in 2008 was part of an ongoing collaboration with HVO to contribute to measuring the gas and particles being pumped out by the volcano, to aid their mission to monitor and understand the volcano and its impacts on the local environment and population.

As is sometimes the case when working on volcanoes, our plans had had to be adapted before we even arrived in Hawaii. Since the early 1980s, the great Halemaʻumaʻu crater at Kīlauea's summit had been quiet. Wafting gas from some clefts and fissures were the main signs that it was still alive. The main volcanic action had been focused to the east, along what is known as the east rift zone, with great lava fields spewing forth from the Puʻu ʻŌʻō vent, a ten-kilometre hike from the Chain of Craters road that snakes through the national park. It was here that we had intended to work that summer. But a sequence of events set in motion on Father's Day in June 2007 – coincidently within hours of the birth of my first child on the other side of the planet – was to change all this. This cycle of new activity started with earthquake swarms, intrusion and eruption near the old Nāpau crater and culminated in the explosive opening of a new vent in the lower eastern wall of Halemaʻumaʻu the following spring. This changed our mission in summer 2008 from studying the Puʻu ʻŌʻō vent to working with HVO on the new gas emissions from the volcano's summit caldera.

My first impression of the area around Kīlauea through the fog of intense jet lag was one of sharp contrasts over a relatively small footprint. The volcano nestles on the south-eastern flank of the great shield-like bulk of Mauna Loa. Massive as Mauna Loa is, there is something about the way its smooth, shallow gradient – a symptom of the relatively fluid basalt flows that built it – recedes gradually up into the hazy atmosphere that seems to underplay the presence its sheer scale might command.

We stayed in a small guest house in the village of Volcano just east of the national park. Here the air felt almost perpetually damp and our cabins were hidden among large ferns and trees. The vegetation was lush and plentiful. In contrast, just 15 kilometres away to the south-west, where we made some of our measurements, is the Kaʻū Desert. Here the dominant trade winds from the north-east have dumped out much of their moisture and blow the volcano's noxious fumes across the land, making the vegetation sparse and scrubby. Sometimes it would be pouring with rain when we left the house but dry and sunny just a short drive away. This combination, with the volcanic emissions in the mix, seemed to mean that rainbows were plentiful; we would often pause from work to admire their colourful arcs across the sky. Sometimes, if we were lucky, they would even span the summit caldera and frame Halemaʻumaʻu's billowing plume.

All the reading that I had done prior to my visit in 2008 had focused on the more recent east rift zone eruption, but activity at Halemaʻumaʻu has been very characteristic of Kīlauea's recorded history. Legend strongly sites the seat of the islands' volcanic power here too. According to Hawaiian fables,

Halemaʻumaʻu crater is the home of the goddess of fire and volcanoes, Pele. Before we commenced our field work in 2008 one of the park rangers performed a ceremony to honour her and ask permission for our endeavours. It was very moving: a moment to reflect on the place and the power of the Earth. I wished that I had brought something meaningful from home to offer to Pele, rather than the hastily plucked, if colourful, flower from the garden at our lodgings that I cast into her crater. I promised myself to do better next time.

Pele's name is part of the global scientific language of volcanology. Pele's hair is the term used to describe the delicate fibrous glass strands extruded as bubbles burst from a lava's surface, frozen and then wafted up in the gases. Her tears are the small shiny pear-like beads made in a similar fashion. During my first trip to Masaya, I gathered great handfuls of these glassy products that had become caught around grass and rocks downwind from the volcano. Its manufacture is like nature's version of rock wool and amassed in the vegetation it looked like a darker, shinier version of the strands of sheep fleece often snagged on barbed wire on country walks back home. These samples later got impounded as animal products by Australian customs on their way for analysis with a colleague in Perth, and I had to write a letter explaining their extraordinary, and most importantly non-biological, origin to get them released. When Masaya's raging magma sea was open to the air in 2017, Pele's hair and tears carpeted the floor of the gassy car park like a steely mass of gold-tinged spiders' webs glinting on grass on a dewy morning. I left footprints in them as I worked.

According to legend, Pele is not originally from the Hawaiian Islands, and there are many versions of the story of her arrival. Apparently fleeing a sibling quarrel, her progress seems to mirror the slow creep of the Pacific tectonic plate, with Pele first attempting to settle on the north-westerly Kauai but then making her way, tracing the geologically ever-more youthful,

The author's footprints in a ground cover of Pele's hair at Masaya's crater rim in 2017

'younging' land to the south-east, to finally make Halema'uma'u her abode at the heart of the current volcanic activity. Pele's capricious nature is revered and respected. She can both destroy homes and forests and create new land where her lava flows cascade into the ocean.

More recently, her ability to bring droves of tourists to the islands, with money to spend, has been celebrated. Other times she is appealed to for mercy. In 1881, lava from Mauna Loa threatened the city of Hilo. Following petitions from the residents, a princess from the royal family in Honolulu arrived, approached the flow front, chanted ancient incantations, offered silk scarves to Pele and doused the lava with brandy. The next morning the lava stalled. This story reminds me of one of my favourite places on the flanks of Etna, where a tiny Catholic chapel sits partly buried in a lava flow just outside the village of Fornazzo. It was apparently built to mark where Etna's 1950 lava flow ended, sparing the village, and then it was partially engulfed by the 1979 flow and is hailed by locals as having played a miraculous role in guarding the village from this flow too. Many other powerful stories of intervention by higher powers permeate the culture of volcanic places like Hawaii and Sicily.

When we worked on Kīlauea in 2008 there were three plumes of gas billowing from the islands resulting from the volcano's endeavours. As well as the emission from the renewed summit activity, the vent at Puʻu ʻŌʻō was still belching out fumes. Further, the volcano's flows pouring into the Pacific were spawning another 'ocean entry' plume. We never made it to Puʻu ʻŌʻō to see the source of its gas, but we felt its touch in the damp acidic haze that blew some days with the trade winds grounding on the Chain of Craters road near where it cuts the contours of the steep cliff or *pali* between the summit caldera and the coast. In Hawaii there is a special term – vog – for this acrid mist or volcanic smog. We sampled its composition by condensing its dew drops on lines of thread strung vertically and hung like a conical spindle from a small tree. We left it dangling above the fern-fronded ropes of old lava overnight with the droplets running down to collect in a bottle attached (with duct tape, of course) to its inverted apex. The 35 grams of clear liquid were a small chemical treasure for analysis.

We witnessed the power of Puʻu ʻŌʻō's activity at the coast, too. While we were there, the molten rock was flowing in channels, deltas and falls down the *pali* to the sea. Sometimes these flows cover themselves in and form tubes hidden by the arch of their own solidified crust. These lava tubes are not uncommon on Hawaii or indeed in other flow fields of runny basaltic lava, and William Ellis records the folklore of the islands, describing how 'Pele went by a road underground from her house in the crater to the shore'. We went to view the ocean entry point at sunset. Where the lava poured into the ocean it flash boiled the seawater, inflating a great cloud of white steamy plume (known locally as 'laze', from lava and haze) – rich in hydrochloric acid from the vaporised waves combining with volcanic gases – over the Pacific. Under the sea, unseen by us, the rock was billowing out in fast-cooled pillows of lava. This combination of hot rock and water is impulsive and explodes unpredictably. We were careful not to approach too close and sat and watched as the daylight ebbed and the white curtain of the plume darkened but for a pulsing iridescent orange fringe at its base where its droplets caught the lava's glow in the twilight gloam.

This process of lava pouring into the ocean builds as well as explodes. New land will rise in places from the sea as layer upon layer of hot rock cools and freezes. Since the 1980s, several square kilometres have been added to the island. It is not immediately prime real estate with its desolate surface resembling something like a frozen dark sea in turmoil. Sharp and rubbly 'a'ā lava interleaves with freer flowing, smoother *pāhoehoe* textures – more Hawaiian words that have been branded into the global language of volcanology. Freshly frozen lava surfaces like these are endlessly intricate, with undulating and twisty ropey surfaces, petrified waves and outbursts. Sometimes in the patterns of *pāhoehoe* lava I can trace a hint of the swirling puddles that I carve in the water with my canoe paddle back at home, and this seems particularly apt for land born from the Pacific waves.

But our main mission was to study the dense plume coming from the recently opened vent within the volcano's summit caldera. Kīlauea's plumbing system has been intensely studied by HVO and other scientists over the years. In the 1980s and 1990s new techniques allowed earthquakes to be located stretching up from 40–50 kilometres beneath Halema'uma'u's crater and spreading at depths of a few kilometres to the east and south-west. These inner vibrations of the planet track the magma's movement as it wells up from the mantle. They tell of it staging about three kilometres beneath Pele's home at the summit and then dyking out horizontally through the shallow crust to break forth in east rift zone eruptions like Puʻu ʻŌʻō.

This seismic tale of the subterranean journey of the magma was corroborated by the gas emissions too. Unlike the carbon dioxide bubbles in champagne or soda, as we have seen, gases dissolved in a magma are a cocktail of different chemical elements and compounds including water, carbon dioxide, sulphurous gases and hydrochloric acid. Each of this menu of gas types tangle up into the structure of the molten rock in different ways and their varying modes of entanglement mean that they escape the gas under different conditions. In the late twentieth century, experiments were done to measure how much of the various volcanic emanations could be squeezed into small packets of molten rock of different chemistries at high temperature and pressure within laboratory furnaces. These miniature simulations of magma's journey to the surface of the Earth yielded many useful lessons. We get insights from nature, too, in the form of microscopic beads of glass – melt inclusions – which are captured, trapped and frozen within a magma's crystals as they form, thus preserving the deep liquid's compositions for us to measure in the lab. Both experiments and measurements showed how magmas with higher concentrations of silica, that is the stickier, more viscous types of magma, can hold more water dissolved in their rocky liquids – the volcanic equivalent of more ordnance to detonate, another reason that they are more explosive.

From such studies we have learnt, too, that carbon dioxide is pushed out of the magma at greater pressure than other gases. This means that bubbles of carbon dioxide fizz into being within rocky melts deeper within the Earth than steam, sulphur gases or hydrochloric acid. Once these carbon dioxide bubbles form, they can work their way out through fractures in the crust and bring messages from lower reaches within a volcano's plumbing.

Where this happens, we can trace the whispers of this hidden chemistry playing out in the gases that we measure sitting as we are on our planet's surface, high above the magma's subterranean flow. In recent years at Etna, scientists have measured faint bursts of carbon dioxide pulsing through the volcano, heralding the arrival of fresh magma deep within its recesses and putting them on their guard. At Kīlauea prior to 2008 and the opening of the new summit vent, the carbon dioxide would leak out of the magma stored beneath Halemaʻumaʻu and seep invisibly into the atmosphere through a broad area of the summit caldera's floor. Other volatiles, like the sulphur gases, would remain held in the magma beneath Kīlauea's summit and then travel underground with its flow to Puʻu ʻŌʻō to be exhaled with the lavas pouring forth there. The opening of the new vent in 2008 led to the question of whether this picture of the subterranean workings of the volcano had changed. Jaggar's measurements of the gas from the 1917 summit lava lake meant that there was the potential for a direct comparison of the volcano's gas plumbing almost a century apart.

As at Masaya, we were able to drive into the dense volcanic gas and it was a relief not to have to haul car batteries and heavy pumps up a mountain. Due to our proximity to the open vent and the high gas flux the plume from Halemaʻumaʻu was stronger than most I had experienced before. Even through my gas mask and goggles, great whiffs of it would occasionally break through, causing my eyes to stream and my lungs to rasp. We would set our equipment running and make a swift retreat. Even in the heat of the day the white cloud was relentlessly opaque and we could not see what was concealed within the volcano's mouth. Clues were revealed at night with the glow of lava illuminating the plume from below. Had I been able to return over the years following our trip I would have watched the crater collapse to frame a lava lake yielding a view much more akin to that which greeted Thomas and Helen Jaggar in 1909. Intriguingly, despite this superficial similarity, our gas measurements showed that something was also profoundly different in the early twenty-first-century lava plumbing compared to that of the early 1900s. The gases contained significantly less carbon dioxide than those recorded in 1917 by Jaggar. This recent summit lava lake, like Puʻu ʻŌʻō, was being fed by magma that had already shed much of its carbon. The details of how magma is connected in the shallow spaces within a volcano are hard to decipher, but this lower gas flux suggested a lower risk of explosions, at least one driven by the magma on its own rather than its interaction with local groundwater. Although cryptic, these murmurs hidden in Pele's breath gave us clues about her shifting mood.

But the reasons for sitting in Halemaʻumaʻu's choking plume that summer were not just about looking into the Earth to understand what volatiles it holds and what the gases tell us about the volcano underground. As we have already touched on, for water, long-term volcanic outgassing has had, and continues to have, a profound effect on our planet's environment and its evolution on geological timescales. We will discover more about these longer-term volcanic effects in our adventures to come. That said, despite their importance for understanding our planet's broader past, present and future, these long-term effects are hard for us to feel and relate to on the scale of our day-to-day experience or brief human lives, and there are far more immediate ways that we can experience the wider reach of volcanic gases. Back in 2008 we were there to look outwards, with a mission to understand the impacts of the volcanic gases and particles on local people living downwind from the new vent, as well as inwards into Kīlauea's magmatic plumbing.

While we were there making measurements at the crater rim, another team was working to launch an instrumented balloon into the plume. The idea was for this gently buoyant device to drift downwind with the gases and particles and trace their path as they mixed and mingled with the ambient air, monitoring the changes in both the plume and the atmosphere induced by this blending. The combination of volcanic fumes with the thin blue line of our atmosphere can change our environment. The atmospheric effects of volcanoes take many forms and take place over a range of scales both in space and in time. They are perhaps easiest to spot when they are relatively swift in their onset and their reach is global. Having adventured underground, we now cast our eye outwards to think about how these forces of nature, rooted deep within our rocky planet, can reach out and touch even the corners of the planet most remote from volcanoes on a very human timescale. With a certain type and size of eruption, we can all feel, at least temporarily, much more acutely aware that we are all inhabitants of Volcanoland.

Chapter 5

Volcanic Winter

*How can big volcanic eruptions
briefly chill the planet?*

I n August 1883, the small volcanic island of Krakatau (also known as Krakatoa), nestled in the Sunda Strait between Sumatra and Java in Indonesia, blew itself to pieces in truly cataclysmic fashion. During its climax the volcano blasted out about 20 cubic kilometres of hot rock, mainly as pyroclastic flows that swept across the straits, triggering volcanic tsunamis and tragically killing about 36,000 people. Like the more ancient and larger Bronze-Age Minoan eruption on Santorini, Krakatau is sited on one of Earth's subduction zones where magmas are beaded from the mantle as one tectonic plate sinks beneath another. As at Santorini, this huge Indonesian eruption was born of sticky, silica-rich magma. And, as with the Minoan eruption, Krakatau devastated the landscape in and around the Sunda Strait. The charts for this important artery of commerce had to be re-drawn. Two-thirds of the island of Krakatau was destroyed and the eruption sank a submarine caldera 250 metres into the sea floor.

In the more than three millennia between the Bronze-Age Minoan eruption and 1883, a heartbeat for our planet, human civilisation had, of course, transformed. While the tectonic forces that shaped the volcanic paroxysms in Greece and Indonesia were defiantly indifferent to the advance of human society, the technological backdrop to this late-nineteenth-century eruption profoundly changed how our species understood this event. Thanks, largely, to the newly laid under-sea telegraph cables, Krakatau's eruption was the first to make the headlines across the world within hours of its unfolding. The eruption became a global event and this rapid knowledge of it allowed observers around the globe to link what might have seemed like disparate phenomena to this distant devastation. Tragically, much was destroyed that fateful August in Java, but it also opened a new chapter in our perception of volcanoes as a truly planetary-scale process. Krakatau's effects were felt in different ways across the world, opening up questions about how a short-lived volcanic eruption can change our atmosphere and environment, over what stretch of the Earth and for how long. Questions that we are still trying to fully fathom today. Before 1883, the boundaries of Volcanoland's influence likely felt remote to many, confined to the landscapes proximal where magmas breached the crust. The events at Krakatau set the scene for a wider appreciation of the global reach of large vol-canic events.

There were many ways that Krakatau's tendrils stretched around the globe. Great rafts of floating pumice, born in the volcanic violence, drifted slowly on the ocean waves away from Indonesia. Some washed up on the shores of islands in the Indian Ocean and the coast of south-east Africa over the following year. Accounts tell of bobbing rocks bearing the tragic remains of some of the volcano's unfortunate victims. In Zanzibar, human skulls and bones were left abandoned 'all along the beach at high water-mark'. According to the headmistress of a local mission school, these were 'quite clean' and 'had no flesh remaining on them'. I imagine that this ivory tidemark must have glinted ghoulishly under the African sun while the children played games throwing the implausibly buoyant pumice stones into the foaming waves.

The detonations from the eruption are reputed to be the loudest noise ever documented. Its sound shattered eardrums on the British ship RMS *Norham Castle* just 60 kilometres from the volcano and was heard over 850 kilometres away in Singapore – and probably further. Subtler, but longer-lived, the inaudible pressure waves spiked the ink traces of barometers across the planet, transforming them into seismometers of the air. For 15 days, the explosion's energy pulsed round the atmosphere, circling it seven times before fading into the background noise.

But it was not only via ocean-borne pumice blankets carrying macabre cargoes and subtle atmospheric echoes that Krakatau reached out around the globe. While working some years ago on a paper about volcanic lightning, I was lucky enough to get access to an original copy of the 1888 *Report of the Krakatoa Committee of the Royal Society*. The book was old and fragile and had to be handled with care. Its embossed cover was worn and faded and a musty smell emanated from its pages as I opened it. Folding out the first leaves revealed, somewhat to my initial surprise, a series of six vivid chromo-lithographs capturing the stages of a single sunset from the banks of the River Thames in London's Chelsea in November 1883. Sunset was at 3.57 p.m. that day, but William Ascroft's crayon sketches depict in the sky 'the final dying out of the after-glow' at 5.15 p.m. The deep reds grade to bright yellows and then fiery oranges and 'the increase of light after the cessation of ordinary twilight ... is very marked'. These six images are among the 533 that Ascroft created in the months following the eruption, a period in which he became fascinated by the vibrant evening skies.

These extraordinary twilight glows had been noticed and connected to Krakatau by various writers in the British Isles. Elsewhere, others had reported blue and green suns in several tropical countries. In Brooklyn, New York, the volcano's reach impressed 16-year-old Frank Perret as he watched the Sun set 'burnished like a copper ball' due to Krakatau's global stretch, which likely sowed one of the seeds of an obsession that would one day lead him to pursue volcanoes across the globe. In Poughkeepsie, north of Brooklyn along the Hudson River, the local fire crew was dispatched in response to the 'light of an immense conflagration' that turned out instead to be the vivid colours of the setting Sun. The wry report in the local paper notes that '[t]he sun is rather a large customer to tackle in the ways of fire, it beats a barn to death. We think the boys did well not to run out of hose.'

By Ascroft's reckoning, these dramatic skies did not entirely fade from view until the early part of 1886. These celestial phenomena became synonymous with the eruption. 'Krakatoa sunsets' were noted by Tempest Anderson as observed in Europe in the autumn of 1902 and he believed that there was 'no reasonable doubt that they were due to Santa María', whose eruption had devastated the Guatemalan landscape far away across the planet. We continue to be intrigued by the impacts on the human psyche of these post Krakatau skies. In 2004, an astronomy professor revisited the circumstances surrounding Edvard Munch's painting *The Scream* and proposed that the picture's famous backdrop of the sky with 'clouds like blood and tongues of fire' was inspired by a post-Krakatau sunset. This remains speculation, but Munch's journal entry recording the inspiration for the painting while walking near Oslo at sunset as the feeling of 'a great unending scream piercing through nature', chimes, likely by chance, with the other phenomena of sound and suffering that attended the eruption.

We are still wont to notice these celestial light shows after eruptions today. Following the eruption of Alaska's Kasatochi volcano in the summer of 2008, the internet buzzed with reports of unusually fiery orange sunsets and ruby red rings around the planet Venus. Around this time I was at a wedding in East Portlemouth in Devon, and as the Sun set over the Salcombe harbour I convinced myself and other guests that the flamboyant colours of the glowing sky were painted for us by the volcano. It is hard to know if this claim was scientifically accurate, but it was certainly a good conversation starter.

Atmospheric shock waves, fiery sunsets and pumice rafts, however morbid the cargo, while intriguing might not seem like truly world-shaking phenomena in terms of a volcano's global reach. But hidden in the vivid colours of Ascroft's twilight skies are deeper, more profound messages about how volcanic eruptions alter our atmosphere and its energy balance. Back in 1883, the reporter in Poughkeepsie attributed the peculiar light in the sky that had launched the zealous firefighters into action as caused by the 'reflection of the rays of the declining sun upon the haze in the horizon'. This explanation captures the essence of the science of these sunsets, although a fuller explanation would augment the more familiar concept of 'reflection' with the more esoteric process of 'scattering', whereby the direction of light is changed as it hits a molecule or particle in the atmosphere. We are doused in scattered photons from the Sun every time we go outdoors into daylight. Scattering is the reason the sky is blue; in fact, it is the reason that the sky has any colour at all rather than being the black backdrop that we see in pictures taken on the surface of the Moon. Earth, unlike the Moon, has a significant atmosphere, and the air molecules scatter the light from the Sun as it passes through, directing it off at oblique angles and onto our retina as we gaze at the heavens. This type of scattering is more efficient the closer in size the wavelength of light is to the air molecules themselves. Air molecules are about a thousand times smaller than the wavelengths of visible light but the blue and violet bands at one end of the rainbow of outputs, held within each sunbeam of white light, have shorter wavelengths than the reds and oranges. This means that blue and violet photons get bent out of their direct line more effectively so that these wavelengths are more likely to reach us as we turn our eyes to a patch of clear sky making it look blue. 'Enough blue sky to make a pair of sailor's trousers' as my grandmother would say as the clouds cleared on a rainy day and we could go out.

At sunset, when the Sun's rays stretch out to us at a slant and take a much longer journey through our atmosphere, we increasingly lose the blue end of the spectrum as it is scattered away before it reaches us. Dust or other particles lofting in the heavens scatter light, too, amplifying the hues. Sitting watching the Sun sink over the Thames in Oxford, I like the idea that the blue light from the fiery orange beams hitting me might be lifting the hearts of friends in places like New York or Boston as they gaze up at the sky. A red sunset is part of the explanation for why the sky is blue. Ruddy sunset colours are the glowing embers of the daytime, although in truth they are the smouldering remains from someone else's daylight and from a day that is still being lived somewhere to the west.

The longer the journey of the light beams, the more of the shorter wavelengths are scattered away and the redder the light becomes. Sometimes over the sea in Devon the Sun can set so large and red that it feels as if it should swallow the waves and take me bobbing in my kayak down with it. After Krakatau, the upper atmosphere, more specifically the stratosphere, had been sullied by the volcano's gas and particles thrust up through the lower atmosphere to an altitude of over 25 kilometres by the volcano's energy. This high in the atmosphere the cocktail of volcanic emissions and their chemistry enhanced the scattering properties of the atmosphere, accentuating its colour palette and bathing the world in the 'Krakatoa sunsets' that so intrigued Perret and agitated Poughkeepsie's firefighters.

This enhanced stratospheric scattering also kept the skies luminous after the normal limits of dusk. High in the atmosphere Krakatau's emissions acted to bounce the light of the sunken Sun back over the horizon even after it had set, illuminating the now shadowed lower atmosphere (the troposphere) and observers like Ascroft frantically painting by the Thames. His depiction of this post-sunset light or 'afterglow' captured by him at 5.15 p.m. that November night in 1883 was one of the things that so struck me, as it has others, when I first found it preserved in the Royal Society volume: an echo of the Sun's orb dropped beneath the western horizon more than an hour since, but mirrored into his consciousness by the volcano's veil of stratospheric haze.

But this interaction between volcanic haze in the stratosphere and the Sun's radiation has effects beyond pretty colours in the evening sky. We bask not just in the light but also the heat of the Sun driving the climate and hydrological cycle and making our, as far as we know, unique living world possible at all. Perhaps surprisingly, given the assiduous cataloguing of Krakatau's reach in terms of barometric pressure and fiery skies, scientists in the 1880s did not immediately think to amass the global records of temperature after the eruption. When the data were finally compiled it showed a drop in temperature worldwide, averaging about 0.5°C in the northern hemisphere. Despite Krakatau's infamy and importance as our society's first global volcanic media event, this was neither the first time that this far-reaching eruptive chill had been noticed, nor is it the most extreme example recorded by science.

As far back as the days of the Roman Empire the influence of volcanic eruptions on climate has been the subject of conjecture. The Greek historian Plutarch wrote of the haze from the 44 BC eruption of Etna rendering the rays of the sun so faint that fruit could not ripen and withered on the trees in the chilly temperatures. In Rome the diminished sunlight coincided with the assassination of Julius Caesar and the eruption apparently became entwined with the emperor's death in the minds of many Romans. Scrolling forward almost two millennia, the American polymath Benjamin Franklin is credited with the first scientific measurements of this perhaps seemingly paradoxical link between the heat of a volcanic eruption and a subsequent cooling of the climate. In 1783 he was living in France as the American ambassador and became fascinated by the strange weather that year. A dry haze had settled over Europe that summer, dimming the Sun's rays such that 'when collected in the focus of a burning glass, they could scarce kindle brown paper.' The planet's 'surface was early frozen' that year and 'perhaps the winter of 1783–4 was more severe than any that had happened for many years.' Franklin, and other scholars bathed in the same haze, posed the hypothesis that its source was a volcano in Iceland. As we will see, the volcano in question was the Laki fissure event that tore into the volcanic island for several months between 1783 and 1784.

Strange weather came again in the years following 1815. The summer of 1816 across New England, Canada and western Europe was exceptionally cold. Records from Yale in the US, the Lancashire plain in England and Geneva in Switzerland noted the summer as the coldest since records began. Snow fell in New England in June and the year became enshrined there in folklore as 'Eighteen Hundred and Froze to Death'. Many emigrated towards the western states. Crops failed and food shortages were widespread. Rioting broke out in France. In Switzerland there were reports of all kinds of things being eaten: from moss to cat flesh. In the summer of 1816, Lord Byron, Percy Bysshe Shelley and Mary Godwin (later known as Mary Shelley) holidayed together on the shores of Lake Geneva and were kept indoors by the incessant rain. They amused themselves with ghost stories and it was under the influence of this grim weather that Mary conceived her gothic masterpiece *Frankenstein* and Byron his apocalyptic poem 'Darkness': 'Morn came and went – and came, and brought no day'.

There is little doubt that the weather in 1816 was widely seen as exceptional. The American statesman Thomas Jefferson wrote to an old friend in the September, 'We have had the most extraordinary year of drought and cold ever known in the history of America'. It has since become known as the 'year without a summer'. Thinkers floundered for explanations. Sunspots, Arctic ice in the North Atlantic and, ironically, given that his work had hinted at what we now believe to be the true cause, the lightning rods invented by Franklin were all proposed. Despite the juxtaposition in many newspapers of the reports of the bad weather with tales of large floating pumice islands in the Indian Ocean, no one joined the dots to deduce the true cause. A world away from Godwin, Shelley and Byron holed up in the gloom on Lake Geneva, in April 1815 Mount Tambora volcano on the island of Sumbawa, in what is now Indonesia, had blown in the largest known explosive eruption of the last 500 years, killing at least 71,000 people.

Despite being of significantly greater size and causing many more fatalities than Krakatau 68 years later, this colossal eruption of Tambora does not seem to have entered global consciousness in the same way. This is likely because in 1815 news only travelled as fast as ships could carry letters. The distractions of other major events, like the conclusion of the Napoleonic Wars, playing out at the time also probably contributed. Maybe, given the clearer disruption to global weather of this earlier eruption, if the dates of Tambora and Krakatau had been reversed, the link between volcanoes and climate would have been the focus of a Royal Society volume rather than the emphasis on sunsets and pressure waves in that about Krakatau. As Earth scientists we are often at the whim of the events that our planet throws our way. We can only study the 'live' events that nature provides during our geologically brief human lifespans, and we can only work with the technology at our disposal at the time. Unravelling the mechanisms by which volcanoes interact with Earth's climate to induce a volcanic winter in summer like after Tambora had to wait until we had taken many more steps in terms of instrumenting, measuring and understanding our planet as a whole. It requires that we wind the clock forward from these great eruptions of the eighteenth and nineteenth centuries to a time when science has compiled eruption records across the world much more widely, when we have bored deep holes into polar glaciers to peer into our planet's past, when aircraft routinely ascend through our atmosphere's tiers, when computer models allow us to disentangle the chaotic trace of cause and effects through the dizzying complexity of the climate system, and when constellations of satellites circle the planet returning great swathes of data that allow us routinely to see our world as never before.

Observations like those recorded by Plutarch and Franklin placed volcanic haze firmly at the root of the subsequent climatic and atmospheric phenomena. Working out the specific chemistry of this sun-dimming mist has been key to understanding how volcanoes can chill the climate. When scientists started to re-visit the link between volcanic eruptions and cooling in the first half of the twentieth century, their assessments assumed that this haze was largely formed of the silicate dust of fine volcanic ash. The direct sampling by aircraft in 1963 of the particles in the stratosphere after the eruption of Indonesia's Mount Agung showed otherwise. These measurements revealed that rather than finely pulverised rock, the volcanic haze was principally composed of sulphuric acid. Ash, even the finest wisps of it, settles out from the atmosphere just too rapidly to account for weather perturbations lasting seasons or years.

Since the second half of the twentieth century, evidence for volcanic sulphur's global reach has also been found in ice cores. These stretched cylinders of solid water are drilled out of the world's persistent ice caps in places like Antarctica and Greenland. Here snowfall accumulates year after year, gradually compressing into thin annual layers of glacial ice. Pressed into these translucent leaves of compacted snow is a record, a memory, of millennia upon millennia of Earth's history. Information about our planet's past is trapped in the ice's tiny bubbles, coded into the impurities clutched among its lattice and even written into subtle changes in the masses – the isotopic balances – of the water molecules themselves.

At the British Antarctic Survey (BAS) in Cambridge I was once lucky enough to be shown a plug of ice that fell as snow an estimated 140,000 years ago. It was carefully shrouded in plastic and lying in state in a polystyrene sarcophagus in their cold store, ready for study. One of the scientists held up a truncated disc of less precious ice to show us. The tiny trapped bubbles made it look like frosted glass. Hidden among the ice's matrix in cores like these are the ancient chemical echoes of volcanic explosions, sometimes otherwise lost to geology. In some cases these take the form of tiny ash shards, but more ubiquitously their tells are spikes on our analysers showing unusual increases in the sulphate content and acidity of the ice. These translucent slices of chilled time are the final icy resting places of sulphur belched and blasted from the inner Earth in a moment of past geological tumult. The ice corresponding to the years after Tambora's eruption in 1815 shows a hefty peak in sulphate and acidity in many cores from around the globe, the settled remnants of the great blanket of volcanic sulphate aerosol that veiled the planet after the eruption. Signals from many volcanic cataclysms can be found in these cores; tie points to particular moments in time marked through their archives. For some eruptions known to human history, like Tambora and Krakatau, the spikes on our graphs of varying sulphate concentration with increasing ice depth are complex epitaphs to the volcano's victims. The source volcanoes for some peaks still remain mysterious. But, especially where they show in the ice from both the northern and southern hemispheres, we can imagine our forebears shivering in their aftermath. This is not hard to do in the chill of the BAS cold store.

Sulphate aerosol from volcanoes and its effects is something that I have studied on and off in different ways for over two decades now. I have already described the haziness of midday on Masaya's crater rim in Nicaragua and the opaque clawing moisture of the dense cloud emitting from Halemaʻumaʻu during my adventures sampling in their plumes. Measurements made by me and others have shown that even straight out of the seeping mouths of these volcanoes, this haze is largely made up of tiny sulphuric acid droplets – aerosol – suspended in the flow of the other volcanic gases. It is evident from its very appearance as mirage-like haze or at other times its more cloud-like opacity that this aerosol is interacting and modifying sunlight, termed scattering as discussed earlier. We can perceive much about scattering with just the naked eye, for example, by noting how far we can see distinctly and how clear objects are in fine weather compared to through a mist. We know that on a hazy day we will lose sight of hills on a distant horizon or perhaps, in more extreme fog, our hands extended a mere arm's length from our face. But to understand the interaction between light and aerosol more precisely we need to put these changes into numbers. We need to measure it in a way that can be compared with our theories and calculations and this means going beyond relying solely on human perception and instead quantifying the power of light via some objective scale. Franklin used evidence from the unusual inability of sunlight to kindle paper. These days we use electronics to count the photons of light that make it along a particular pathway through the atmosphere to hit a detector.

Understanding the properties and effects of the haze emitted from volcanoes is an important component of comprehending the phenomena noted by Franklin in 1783 and by others after Etna, Tambora and Krakatau. The global haziness of the atmosphere, or aerosol optical depth, to give it its more scientific term, has been systematically monitored for over 25 years now by a network of instruments known as sun photometers trained on the heavens. We can also take smaller portable versions of these instruments and hike them up and around fuming volcanoes to take a closer look at how their aerosol plumes are changing the Sun's rays. This is what took me to Chile in 2003 for my first encounter with the volcanoes of this long, thin country fenced in as it is by the Pacific, the Andes and the Southern Ocean. In her memoir *My Invented Country*, which was first published a little after my first visit, the Chilean writer Isabel Allende describes the country as 'a

Villarrica volcano from Pucon, taken in 2003

lance to the south of the south of America'. 'Shaken by the sighs of hundreds of volcanoes', it is a wonderful place for volcanic adventures and somewhere I am perpetually drawn back to studying.

I sometimes get asked which is my favourite volcano. This feels like an impossible question, but if pushed for an answer, to keep the conversation going, I have been known to say Villarrica in south-central Chile. South America's crooked backbone of the Andes is lower here than in the north of the country. The climate is wet and the vegetation lush. In *My Invented Country*, Allende describes the area around Villarrica (or Rucapillán, meaning 'great spirit's house' in the indigenous Mapuche language) as 'an enchanted region of forests, lakes, rivers and volcanoes.' When I first arrived in the area that February in 2003, the cloud deck was low and rain was flowing freely from the heavens. As Allende says, '[r]ain and more rain nourishes the tangled vegetation of the cool forests where our native trees rise tall.' Without sight of the Andes Cordillera to anchor me, it reminded me strongly of the British Lake District, also known for frequent rain. In both places in the torrents, the landscape feels urgent with water as it tumbles from the sky and down the gradients. As in these British mountains, during downpours here, waterfalls reveal themselves from more subtle latency in the landscape, opening like white gills flared by the weather or torn curtains hanging from the cloud base against the green slopes. Strangely for someone born and raised in the damp of the British Isles, having come on that particular trip from three weeks in the Atacama Desert, I felt a heightened awareness of the frenetic lines of water hammering the roofs and landscape, and the low clouds initially gave me an unaccustomed sense of claustrophobia.

The eyes of geologists seeking to make sense of the world have traced this landscape for centuries. Charles Darwin visited the region in the austral summer of 1834–35 during his voyage on the *Beagle*. His diaries chronicle the terrible aftermath of the February 1835 Concepción earthquake, local tales of witchcraft stopping a volcano's activity and an account of a small night-time eruption of Osorno volcano. This juxtaposition of close experience of both earthquakes and volcanoes – 'parts of one of the greatest phenomena to which this world is subject' – set him thinking about grand-scale geological theories, but he was also far from immune to the splendour of the landscape. Although Darwin notes casting his eye over the 'generally active' volcano of Villarrica, while he was there it 'appeared quite tranquil'. It is his description of Osorno, 'this most beautiful mountain, formed like a perfect cone & white with snow', that chimes more with my sense of Villarrica's beauty, but in truth this description could be equally applied to several of the majestic peaks that grace this landscape.

Villarrica is certainly a very attractive mountain. The symmetry of its broad cone is breathtaking from many viewpoints, and its snow cap waxes and wanes with the seasons, sending icy white fingers down its flanks. But it is not just a contender as my favourite volcano because of its beauty. The reason is really the story of a moment, a very surreal moment, in 2003. When we arrived in Pucón, the town at the base of the mountain, accommodation turned out to be hard to come by. The rain seemed relentless and we had been warned that weather like this could set in for weeks in these parts. We would not be able to work on the volcano in these conditions so, as we wandered through the town searching for a place to stay, it was hard not to feel low. After a night in a hotel that was strictly beyond our budget, things began to look up. We found affordable accommodation in a log cabin within a camp site off the road on the way up the volcano and the person who took the money from us invited us to a barbecue that evening. The rain had also begun to ease and while there was still no sign of the mountain between the clouds, at least we were no longer homeless and had the prospect of local company to look forward to.

Although it was still the summer, the rain had rendered the evening air cool and the smell of wood smoke curling up from the chimneys of the cabins' stoves hung in the air as we walked down to find the party. Two things quickly became apparent. Firstly, a barbecue in Chile often has quite a different meaning from a barbecue back home. We arrived to find an enormous man with a huge knife and a half-drunk bottle of pisco roasting a whole pig on a spit. Secondly, it turned out that 7 p.m. did not mean 7 p.m. on the dot, and our British punctuality meant we were very early. There followed an hour or so of difficult conversation with our new companion, challenging our limited Spanish vocabulary, while he became increasingly drunk. It was during this awkward scene that the volcano decided to make an appearance. In a moment that in my memory takes on an almost biblical aura, the clouds parted and suddenly it was there, far closer than I had imagined, its dark flanks bleeding up into the white ice cap and perfect puffs of cloud drifting away from its summit. In my head, perhaps helped by the pisco, angels trumpeted. 'Villarrica, beautiful,' acknowledged the Chilean chef, waving his huge knife somewhat dangerously in the direction of the view. Shortly afterwards, some other Chileans turned up and the party started properly and went on well into the night. We emerged late the next day to find the place deserted and bearing no trace of the party. It was as if we had dreamed the whole evening. It turned out that that person who had rented us the cabin was not, in fact, the camp site owner, but instead had been left in charge overnight while they were away. We never saw any of our other fellow revellers again. Further, when we came to leave two weeks later, we learnt (admittedly via our rather broken Spanish) that our first night's barbecue had possibly been paid for with our rent paid in full upon arrival, which may not it seemed have made it to the rightful proprietor. These difficulties aside, in better news the weather had lifted and the way was open for our first expedition up the mountain the next day.

Climbing Villarrica is a long trudge up a zigzag path carved into the ice. Villarrica's glacial cap is the remnant of a great ice sheet that covered this area during the last ice age, and it, like many other ice caps and glaciers in the region, has been thinning more rapidly during recent years. On our first ascent we joined a tourist group led by a local guide. It had not been possible to climb for some weeks in the poor weather and so there were many groups ascending that day, trailing up in close-packed lines, kicking footholds into the icy footsteps of those who had gone ahead. We paused where a dark and twisted lava flow broke through the snow. Perched on the sharp stone I gazed down at the frozen tumble of once flowing rock scarring the bright canvas of the ice. Another morning we rested in this spot and admired the flowers of frost that had grown in the night tracing the lava's ropey flow like foam on a petrified ocean. Near the summit the white snow was dusted with volcanic debris, giving another strange wave-like texture where the dark rock spatter picked out the flow and melt cracks of the ice: rock and ice in layers painting their own perpetually reforming landscapes within each other.

Stamping our way above the broken cloud deck, the double peak of Llaima volcano rose into view. As with Masaya, at Villarrica hot magma was not far from the surface. Although the dense fumes made seeing the magma impossible, it made its presence felt, not just in the stinging gas but also the occasional boom of a small explosion followed by the patter of falling ash, small chunks of bright black scoria – the basaltic equivalent of pumice – and delicate glassy grey reticulite (sometimes referred to rather beautifully as thread-lace scoria). Occasionally this light rocky rain would fall on top of us as we worked. We wandered around the crater rim collecting the fallen fragments while the sampler pumps hummed. They were easy to pick out,

Frost on lava taken during a morning ascent of Villarrica in 2003

plum-sized chunks of jet-black treasure glinting fresh in the Sun. I still have some of these fragments and for a while they were a useful aid in school presentations to get the students to consider that a rock could be younger than they were, even if most stone is extremely old.

As well as sampling with pumps and measuring the gas, we were gauging the optical qualities of the particles and droplets drifting in the plume's cloud using a handheld sun photometer. This small box, about the size of a hardback book, needed to be precisely pointed at the Sun and a pin-hole projection of our star's orb aligned into the centre of a small bullseye target on the front of the instrument. It was fiddly and required patience and a steady hand, especially hard in strong mountain-top winds.

On my first attempts I failed to corral the Sun onto the target at all, or tilted the device far too vigorously and watched as the pale-yellow circle swung wildly across the bullseye. As with most things, I improved with practice, but from a distance, it must have looked as if I had taken the curious decision to play one of those infuriating ball-in-a-maze balance games at the top of a volcano. Inside, optics and electronics split the Sun's beams to measure its power in bands of selected wavelengths. If I had been able to see inside the photometer's box and trace the origins of the numbers output into our spreadsheets at the end of the day, I would have seen a measurement in purple light, one bathed in burning orange and then two channels in the invisible stretched wavelengths of the infrared.

Accurate measures of time, date and location in longitude and latitude had to be set before we could get to work. Inside the photometer's processors these data are melded to quantify the amount of atmosphere the Sun's rays have passed through, allowing a calculation of how comparatively transparent or opaque – optically thick – it is from the measurements. Reclining on the rubbly ground at Villarrica's summit, wedged to keep a line of sight on the Sun, I could feel the reduction in the light and heat as denser puffs of plume wafted overhead, sometimes accompanied by a gentle patter of scoria and ash thrown out by a big slug of gas bursting from the invisible lava below. Downloading the data back at our mountain-side cabin, my sensations translated into time-stamped numbers held within the computer's spreadsheets. Dull as these sheets of numbers might first appear, they take on a new complexion when you reflect that each row tells the tale of a journey. Each value in each cell narrates photon pathways that reach out across the eight light minutes of cold void between the Sun and Earth, weave their way through the many layers of atmosphere, and then the particles and gas of Villarrica's plume, to be finally focused onto my detector. The measurements under the plume that day precisely quantified what I could sense in the dimming of the light and heat as puffs of plume passed overhead. Like clouds, Villarrica's plume looks bright and white from above. Sulphate particles from volcanoes behave like imperfect mirrors and some of the Sun's photons are being bounced back upwards towards the black of space. The puffing volcano smudges a border of influence, below which its cloud dims the world and above which it brightens it.

Villarrica's volcanic plume is very limited in its footprint and its impact on our planet's heat balance. There are very clear differences in scale between the gentle puffing of this snow-capped cone and the epic violence of Indonesia's peaks in 1815 or 1883. But the sheer amount of volcanic material put out is not the whole story. Behind the multihued drama of Ascroft's obsession with the sky's evening colours from 1883 into 1886 recorded in his many artworks is a lesson about the Earth's atmospheric structure. For a short-lived volcanic explosion to paint itself into our skies for years like Krakatau or rob a population of a summer like Tambora, the volcano's outputs have to persist in the atmosphere. This generally requires a high Plinian eruption column with the thermal forces driving it to great altitudes within the heavens. Estimates of the Krakatau eruption height put it over 25 kilometres in altitude. Some estimates reckon Tambora's at over 40 kilometres.

But it is not just a case of what goes up further, takes longer to come back down. The Earth's atmosphere changes as you rise through it, and it is arranged into ascending tiers or, more correctly, concentric spheres. Our lives play out in the basement troposphere. Warmed by the Earth's surface below, the air in the troposphere is constantly on the move at the whim of convection and in dialogue with Earth's surface friction. Tropospheric air is also in constant exchange with the oceans, sucking up water by evaporation to make clouds and weather.

For volcanic gases and particles, all this movement and flushing with water tends to clean out anything remotely sticky, soluble or reactive with water. This means that gases like sulphur dioxide tend to be rinsed out of the troposphere over mere days or weeks. The sulphur dioxide and sulphate haze that I studied escaping from Villarrica's summit into our troposphere will not stay in the atmosphere long enough to reach across the globe, but will instead be pulled down to Earth by the wind and weather in the region of forests, lakes, mountains, pampas and oceans downwind from its majestic peak.

But as you ascend above about 12 kilometres (higher at the equator and lower at the poles) through the ever-expanding shells of thinning air, you enter a new domain. Here in the stratosphere, the thermal architecture of the atmosphere changes. At this loftiness, the Sun's rays contain high-energy ultraviolet photons that hit oxygen molecules, breaking them apart and freeing two oxygen atoms to each hook up with another oxygen molecule and make a triple oxygen molecule known as ozone. More ultraviolet radiation can then split these ozone molecules again and this constant cycling of ozone creation and destruction alchemises ultraviolet radiation into heat energy. This cycling process means that the stratosphere is heated from above by absorbing the Sun's rays. Layers of warmer gas therefore sit above cooler gas below. Unlike the convection raging in the troposphere, this arrangement makes it stable (or stratified), and vertical mixing between these layered shells of air is more sluggish. Since air rising into the stratosphere from below must pass through the cold upper reaches of the troposphere, freeze-drying out most of its moisture, the stratosphere is also dry and largely devoid of weather.

The frantic exchange of photons and electrons that thrums above our heads daily, cycling ozone, cuts out the harmful ultraviolet rays from the Sun before they reach us. In the 1980s, scientists from BAS, among others, found a rapid decline in the ozone above Antarctica that was subsequently shown to be caused by human emissions of halogen-bearing chemicals, CFCs, used in spray cans and as refrigerants. Despite successful mitigation, this ozone hole is only recently starting to show signs of recovery. The relatively dry and stable structure of the stratosphere means that once gases like CFCs are up there, it takes a long time for them to clear out. The same is true for volcanic gas. When volcanoes like Krakatau and Tambora punch their plumes high up into the stratosphere, their exhalations take far longer to dissipate than their tropospheric counterparts. Interleaved within the stratosphere's stable gassy strata, the volcano's sulphur gases can slowly oxidise, creating and replenishing the fine haze of particle-forming sulphate aerosol. Although these tiny particles slowly grow and settle back down through the atmosphere's levels, the lofty clouds from eruptions like Tambora and Krakatau still have ample time to spread around the globe, reaching out to inspire Byron's pen and Ascroft's brush.

Sometimes, when studying the stratospheric effects of volcanic eruptions, I feel sandwiched between the very deep and the very high. We must stretch our minds between the unimaginably crushing depths where the eruption's magmas were born, and the rarefied reaches of the higher atmosphere where its effects play out. A few years back I watched a documentary film, *Miracle in the Storm*, about Ewa Wiśnierska, a German paraglider who in 2007 got sucked into the updraft of a powerful cumulonimbus thundercloud. After a terrifying journey hurtling upwards through the pitch black, hail and lightning of the storm she shot out of its top at an altitude of almost ten kilometres. At the time, I was collaborating on a paper working to measure the height achieved by sulphur dioxide during explosive eruptions using satellite data. Unconscious as she was, according to her flight log, Ewa spent 45 minutes carving out broad circles in the calm above the thundercloud. Although she was technically only at the edge of the stratosphere, her exit from terrifying tumult into the calm above gave me a human reference for the path of the sulphur that I was studying and its transforming circumstances within explosive eruptions. After the violent rise of the plume, dark and dirty with ash and laced with lightning, using satellites we can follow the trail of a volcano's sulphur dioxide as its filaments and tendrils stretch silently through the bright calm of the stratosphere or upper troposphere. It is inconceivable to survive a similar journey through an explosive volcanic plume, so perhaps Ewa's terrifying experience is the closest we can ever imagine to such a ride. Occasionally volcanic plumes or aerosol veils have been intercepted and sampled by aircraft or instrumented balloons, but these encounters are infrequent and, of course, have their own risks. They also only intersect small sections of the spreading volcanic cloud. To really understand how a volcano can have worldwide reach requires scientific measurements and computer simulations that can help us to gain a truly global viewpoint.

We saw how the advent of global telegraph communications changed the way that humanity understood Krakatau compared to Tambora. Since the launch of *Sputnik* in 1957, our planet has been watched by an ever-increasing network of orbiting satellites circling hundreds to tens of thousands of kilometres above us. Some are tasked with peering down at the planet looking for trends and changes. These wide-angle vistas have given us new understanding of our home, revealing connections on a planetary scale in a new way. These eyes in the sky have given us new perspectives on volcanoes too. From outside the atmosphere, we can track the clouds from big eruptions as they bleed through the global stratosphere, as well as logging the climate consequences at a resolution and on a scale unprecedented in the history of human invention.

A step change in our understanding was unlocked by the satellite observations of the 1991 eruption of Mount Pinatubo in the Philippines, among the largest eruptions in the last 100 years, which blasted gas and ash over 30 kilometres up into the atmosphere. Immediately after the eruption, satellite images showed a fine haze made of tiny droplets of sulphuric acid stretching around the planet's equatorial belly high up in the stratosphere. Promoted by the volcano's tropical position, this veil spread into the atmosphere of the northern and southern hemispheres, smearing towards both poles over the months that followed. The fineness of these aerosols made them especially effective at scattering the shorter wavelengths of radiation bathing Earth's atmosphere from the Sun, including the visible and ultraviolet light. By reflecting some of this energy back out into space, this lofting global mist made the direct sunlight at the surface dimmer.

The 12 June 1991 eruption column from Mount Pinatubo taken from Clark Air Base about 25 kilometres east of the volcano

This change, while imperceptible to our senses, was not immeasurable. Perched almost 3,400 metres high on the flanks of Hawaii's Big Island, where the air is generally cleaner and clearer, the Mauna Loa Observatory has been measuring the power of the sunlight since the late 1950s. A dip in the power of the rays followed the Pinatubo eruption, lasting two to three years, and this drop in the energy making it down to the surface showed up as a fall in global temperatures as well. Against a backdrop of global temperature rise since the early twentieth century, the one to three years after Pinatubo were, on average, 0.2–0.5°C cooler.

Other, more subtle changes were discernible too. The particles in the stratosphere promoted chemical reactions that temporarily depleted our shielding ozone layer. While the surface was cooled, the sun-bathed stratospheric particles absorbed as well as reflected energy and warmed the upper atmosphere, altering its ebbs and flows and causing unexpected effects like warmer winters against the broader cooling. Data hint that the changes in the light and heat reaching the surface changed rainfall patterns and the carbon cycle, with some studies suggesting that the hazier sunlight scattered by Pinatubo's stratospheric veil promoted photosynthesis in some types of global forests, allowing their trees to draw in a deep breath of carbon dioxide.

The scientific community is still unpicking the more nuanced ramifications of the Pinatubo eruption. As well as yielding new insights into the global impacts of volcanism, the post-Pinatubo measurements have allowed scientists to explore how well they understand planetary-scale processes like atmospheric circulation and feedbacks on things like the seasons, weather and vegetation. This volcanic rocking of the climate system was an important test for the global computer models of climate used to forecast what might happen in the future to global temperatures, rainfall and wind patterns. Although it cares nothing for our science, volcanic agency performed an experiment on the Earth system on a massive scale pushing at the bounds of our understanding of its behaviour.

That is not to forget that there was a very human toll for the communities living around Pinatubo. Despite successful forecasting of the eruption and evacuation of tens of thousands of people, there were still more than 840 fatalities, largely as a result of collapsing roofs, disease in temporary housing and lahars. Millions were affected by the eruption and hazards like lahars persisted for years afterwards as river systems cleared the volcanic debris clogging their courses. Although when Pinatubo erupted it gave us new insights into the effects of past volcanic eruptions and an opportunity to learn more about how our atmosphere, oceans and climate work in general, for the hundreds of thousands of people living near the mountain it was a desperate and terrifying event. Unfortunately, with volcanoes we do not get to choose where the next one erupts and it is the Earth that dictates both what science we can do, in terms of, for example, the style and size of the eruptions it delivers, but also, more grimly, where we must focus our efforts to save lives and livelihoods.

There could be moments in the future of human civilisation when understanding the effects of big explosive volcanic eruptions becomes considerably more vital than it has been to date. Pinatubo, Krakatau and Tambora are far from the largest-scale eruptions that our planet can produce. Using techniques like those explored in Chapter 3 for Santorini, volcanologists can put together the anatomy of volcanic eruptions far larger than those captured in our short human historical record. The press often like to refer to the very largest eruptions left in the geological record as coming from 'supervolcanoes'. As a volcanologist I prefer the more tempered term 'magnitude eight eruptions' (on a magnitude scale running from zero to nine with each notch up signalling an increase in the eruption size by a factor of ten times). Roughly speaking, these monstrous events are classified as those that spew out more than 1,000 cubic kilometres of rock onto our planet's surface in an explosive cataclysm. The largest known, from the La Garita Caldera in the USA, erupted about 28 million years ago, leaving 5,000 cubic kilometres of petrified pyroclastic flow deposits straddling the Colorado–New Mexico border: a vast deposit known as the Fish Canyon Tuff. None of the volcanic eruptions that I have discussed so far come close. The mighty Minoan eruption from Santorini or the fearsome 1815 eruption from Tambora only erupted tens to hundreds of cubic kilometres of material (magnitudes six to seven). The most recent example of a magnitude eight is the Oruanui eruption of New Zealand's Taupo volcano, about 26,500 years ago. Other famous examples are colossal eruptions from volcanoes such as Yellowstone in the USA or Toba in Indonesia, but they were all tens of thousands to tens of millions of years ago. These are volcanic catastrophes on a scale unchronicled in our short human history.

The significant global chill documented after Tambora and more recently the subtler effects measured after Pinatubo have led to speculations about prolonged 'volcanic winters' (a term coined in grim analogy with the possible apocalyptic aftermath of a nuclear war) after these magnitude eight scale eruptions. In such a scenario, the planet would be tipped into prolonged glaciation by the dimming of the sunlight tempered by the persistent volcanic haze. For events this far back in time, the stories of the eruptions and their interactions with the complex system of Earth's climate are hard to decipher. Estimating how much sulphur these massive eruptions emitted and how high it reached in the atmosphere are key to predicting their effects but shrouded in uncertainty. To understand their interactions with the atmosphere we must rely on complex models of the Earth system, which, while constantly improving, are still infused with approximations and assumptions largely based on the more recent past. So far these models predict a decade of cooling following a magnitude eight-scale eruption, rather than the much more prolonged winter of a glacial tipping point, but there is still much that is debated and a lot to work out. We do not know where or when, but it is highly likely that one day Earth will again bear witness to an eruption on this scale. Despite my deep scientific investment in the topic, I very sincerely hope that it is somewhere vanishingly distant over our future's uncertain horizon. Fortunately, there are no signs of such unrest, at least at present, although given the limits of our experience we cannot be sure of the exact signals we might see in the run up to such an event.

A likelier scenario in the more immediate future is another eruption of a similar size to that of Pinatubo or maybe, if we are less lucky, Tambora. There are many lessons to learn when planning for this from the Philippines in 1991, where, despite the tragic losses, the successful evacuation is estimated to have saved at least 5,000 lives. With ever more sophisticated sensors gridding the Earth's surface and orbit, watching as our planet changes, we are better primed to study it scientifically than at any previous point in history. Eruptions like that from Tonga's Hunga Tonga–Hunga Ha'apai in 2022 have already hinted at how much more there is for us to understand. This eruption was unusually explosive and water-rich, likely due to its partially submarine nature, and thus offers us an opportunity to understand more about the atmospheric impacts of volcanic water vapour, which are usually harder to ascertain given our atmosphere's large and constantly varying water-load. What new things we will discover in the future about the behaviour of the thin blue line of atmosphere that sustains us, and how volcanoes can change short-term climate, is something I cannot predict. Some might even hope that a volcanic eruption of the scale of Tambora or larger could offer us some respite from the relentless trend of warming, although we should remember that such eruptions can also disrupt important atmospheric circulation and rainfall patterns in unpredictable ways and unless it is remote from large scale populations could easily be a harbinger of tragedy. In any case in terms of the global impacts I, quite possibly naively, hope that should we be dealt another 'year without a summer', we are better prepared than in 1815 and no one needs to explore the gourmet potential of their pets.

While stratospheric aerosol veils and volcanic winter might be the most immediately obvious stuff of planetary epics, there are many ways that less explosive emissions, lower down in the atmosphere, can affect the planet and its climate, environment and biology. Franklin's influential observations of the strange weather and reduced power of the Sun in 1780s Europe highlight (or should that be lowlight?) the importance of understanding the impacts of sulphurous volcanic haze in the lower reaches of the troposphere, as well as the remote stratosphere. As we shall see, some eruptions compensate for lower explosivity and smaller instantaneous gas fluxes with extended longevity.

Standing at the low crater rims of Kīlauea and Masaya, you can watch their plumes drifting off with the breeze. Sometimes they loft into the atmosphere but other times they hug the ground, and such low-level plumes often dole out unpleasant effects to the local people. In Hawaii volcanic haze or vog can envelop portions of the islands under certain combinations of volcanic and meteorological conditions and is analogous to the smogs that characterise urban air pollution. In the settlement of El Panama, perched on the old caldera wall just two kilometres downwind of Masaya, pots and pans quickly rust, roofs must be bound to houses with rope rather than nails, and detergents are ineffective because of the volcanic acids carried in the breeze. There is often at least a whiff of the volcano's acerbic breath in the air and respiratory health and water contamination are problems too. The locals have had to change from farming coffee to more resistant crops such as pineapple and dragon fruit. Those who have lived there for a long time tell of how there used to be trees before Masaya began degassing again in the 1990s.

Despite their proximity to the national park, most of El Panama's inhabitants have never been able to afford the entry fee to go and gaze in at the gaping volcanic mouth whose breath they feel daily. In 2017, when Masaya's lava was high and roiling, there were even small fragments and bundles of black Pele's hair in the dusty gutters outside the breeze-block village school, like miniature dark tumbleweed. We were there that evening to screen a film about the villagers' lives made as part of a collaboration with the local volcano monitoring agency. Earlier, we had bumped up the dirt road with an enormous television screen lashed into the open bed of a pickup truck and with 60+ fried-chicken dinners crammed into the back seats for the El Panama premiere. There was something very moving about watching the room lit up by the raging glow of Masaya's lava lake. In my memory, the children's eyes widened and one of them reached out to touch it, but I cannot be sure that this happened. But it was the village's resilient people and their tales of how the ethereal volcanic emanations have changed the landscape and their lives that are the real stars of this short fragment of film.

Among the most extreme examples recorded in history of a relatively non-explosive volcanic event wreaking havoc on the local and regional environment is the Laki eruption that raged from June 1783 to February 1784 in Iceland. It was the emissions from this prolonged volcanic breakout that reached out across northern Europe to trouble Franklin during his time in France. Laki was a very different type of event from the likes of Tambora or Krakatau. Where those events were violent and acute, Laki's longer-lived activity was less explosive but far more persistent. Over the eight months that its 'fires' lasted, the event poured out about 14 cubic kilometres of lava onto the planet's surface to the west of the Vatnajökull ice cap in a series of lava flows, fissures and fountains.

One of the most complete accounts of the events from Iceland itself is the *Eldrit* (Fire Treatise), a chronicle kept by Jón Steingrímsson, a parish priest in Klaustur village about 30 kilometres south of the eruption fissure. It was on his way to preach on 8 June 1783, Pentecost Sunday that year, that Jón first noticed the ominous black cloud moving in from the north. The cloud cut out the Sun and was accompanied by a drumbeat of earthquakes and cracking sounds. Soon acid rain began to fall, scorching the plants and animals. Ash and Pele's hair – 'blue-black and shiny, as long and thick around as a seal's hair' – showered from the sky, worked by the wind into coils like I saw in Nicaragua.

Just days into the eruption, the Skaftá River, that flows close to the eruption site before cutting a gorge through to the coastal plain near Klaustur, dried up entirely and its canyon became a course for lava rather than water. This lava flowed 'with frightening speed' and breached the banks onto the villagers' meadows. I first visited Iceland on a family holiday long before studying volcanoes had become a passion. Nonetheless, the landscape made an indelible mark on me and among the many photos stuck into my album is one of Laki's *eldhraun* (fire lava) field. Shrouded in dense moss, the lava's undulations looked more like a ruffled green sea to me then than something wrought of stone, with hummocks of rock breaking the surface like the white horses of foam on cresting waves. That summer in 1783, this same rocky mass was terrifying, laying waste to trees, homes, farms and churches as it surged forward. By late July, the lava was threatening Jón's own church in Klaustur. On 20 July, with volcanic thunder clapping so loudly that the steeple bells reverberated in response, Jón delivered what would become known as the *eldmessa* (fire mass). This powerful preaching was credited with stopping the flows short of Klaustur, although by

Jón's own admission it likely had more to do with rivers bloated by heavy rainfall quenching the lava up the valley.

Although the immediate threat of destruction for Klaustur passed, the broader crisis was far from over. In the highlands above the village, the suture in the land through which the fire fountains poured tore further to the north. As autumn began to beckon, lava flows started encroaching onto the coastal plain from the north-east of Klaustur, as well as the south-west, threatening to clasp and trap the people in between. Jón thought each limb of lava to have its own individual stench. In his reckoning the odour of that to the east was of burning wet weeds 'or some such slimy material', and that to the west as if 'burning coal had been doused with urine or another acrid substance.'

To the relief of Klaustur's inhabitants, the flows waned over the winter months and on 7 February 1784 their distant glow was seen for the final time. But the end of the lava's destruction was not the end of the devastation. In Iceland the plumes from this enormous volcanic outpouring had poisoned much of the island with fluoride, ash and acid. Crops failed, livestock died and even the streams and rivers were tainted, purging them of fish. Famine and disease set in, and about a quarter of the island's inhabitants perished. The Icelanders have a word for this period of defining hardship and tragedy that etched itself onto the country's psyche: *móðuharðindin*, 'the hardship of the fog'.

Although violent and persistent, Laki's basaltic fire fountains achieved nothing like the towering altitude of the Plinian columns of ash and gases muscling their way into the stratosphere at Tambora or Krakatau. The lava sprays themselves only reached heights of approximately a kilometre, although the hot gases might have wafted up higher under their thermal buoyancy, sometimes maybe even touching the lower stratosphere. Despite largely remaining in the troposphere, fed as they were by the intense and persistent eruptions in Iceland, the tendrils of this gas and aerosol plume spiralled out, spreading across northern Europe in summer 1783. Weather patterns also played their role and collective testimony from across the continent, including that of Franklin, tells of a caustic dry fog turning the sunlight rusty red, dimming the night-time stars, bleaching vegetation and making breathing difficult, especially for those with heart or lung problems. While the summer was unusually hot, the winter of 1783–4 was uncommonly cold across Europe and north America. Precisely linking cause and effect in terms of the eruption and these machinations of climate is tricky. But the case of Laki does provide strong evidence that it is not just the most explosive volcanic events that can reach out beyond their immediate environs to chill and poison the air and land. Lower-level volcanic plumes can have a big environmental footprint, too, if the emissions are intense and persistent enough. This is something that our adventures will bring us back to later, when we consider how Volcanoland's largest and longest events can be intertwined with the fossil record that charts the evolution of life on our planet in Chapter 8.

Back at the top of Villarrica at the end of a long day of measurements, the Sun would start to dip towards the Pacific as the time drew in for us to head home. After the trudge up the mountain's frozen cap, the trip down was a very different affair and we slid, equipment and all, down chutes hollowed like bobsleigh runs into the ice. As the weather window stretched out and we notched off successful ascents, these ice chutes became ever more polished by the tourist traffic and the ride more hair raising. On the final day of our visit, I managed to give myself the largest bruises of my life, blossoming to the size of my two stretched out hands, by crashing down a huge step carved into a chute.

After one particularly long day, the receding sun tinted the volcano's plume and snow cap pink with its dying rays as we walked back to the car at the top of the road. I watched puffs of particles drift off with the wind and followed them with my eyes as far as I could in the fading light. On another day we drove the sun photometer downwind of the volcano and measured the subtle shifts in the Sun's light when viewed beneath the ageing plume, to tell us more about differences in the average size of the particles with transport and maturity away from their emission. In the damp atmosphere of this part of the Andes, the sulphate particles that had been gasped from the volcano's mouth, were slowly growing as they wafted with the wind, grabbing water molecules from the surrounding atmosphere and tying them into the aerosol. On other occasions, from the top of the volcano, I could watch the plume's smudge mingling into obscurity as it irretrievably mixed with the Chilean air. However, the growing particles gauged from below via my electronic box of tricks told me that this fine haze from the volcano will not *always* disappear. Each tiny particle is a potential seed for a cloud droplet that might eventually obscure a clear sky, colour the last rays of a sunset, or fall as rain and feed the region's precious *Araucaria* forests.

Linking a puff of particles from Villarrica's lava lake directly to the fall of rain through the swirling complexity of our planet's weather is hard. But we can see hints of the trails left in cloudiness and cloud brightness downwind from volcanoes in measurements from satellites. Hundreds of volcanoes like Villarrica, Kīlauea, Masaya, Etna and many more pump out great quantities of gas and particles into our lower atmosphere every day. Although, especially where the plume grounds, the local environment might feel some effects, these volcanoes on their own don't change the Earth dramatically – or at least not in a measurable way. But if we were to turn them all off forever, we would, in the longer-term, begin to realise their impact on our planet. We already touched upon how they are part of the finely poised long-term cycling of water between Earth's interior and exterior that maintains the oceans and oils plate tectonics. But this exchange of water between the inner and outer planet is just one of the many similar planetary-scale chemical rotations in which volcanoes play a role. We have seen here how volcanic sulphur can reach out and rock the Earth's climate system. But the dramatic atmospheric perturbations following eruptions like of Tambora or Toba play out against the backdrop of a far more nurturing role of volcanism for our long-term environment, and the Earth would not be the relatively benign and stable place that it has been over the eons without the everyday action of the fuming peaks in Volcanoland. Carbon is a key driver of our planet's long-term climate. As we saw back with the asphyxiated canines in the Grotta del Cane (recounted in Chapter 4), carbon dioxide is a major component of volcanic emanations, and it is to the – often cryptic – degassing of this potent gas that we now turn. Here again it seems that Earth has been dealt a very fortunate set of circumstances that have blended to make it ultimately able to host the complex variety of life we see around us today. As we shall see, when it comes to preserving climate stability on a habitable planet, it really is all in the balance.

Chapter 6

All in the Balance

How do volcanoes help keep
Earth's long-term climate stable?

In 1984, Haraldur Sigurðsson, an Icelandic volcanologist working at the University of Rhode Island at the time, was dispatched by the US State Department to Africa to investigate the sudden and mysterious deaths of 37 people around Lake Monoun in Cameroon. When Haraldur and his PhD student reached the area, they found that people had 'died in their tracks' as they walked along the lake shore in the predawn hours on their way to market in a nearby village. There were few clues to the cause of these tragic, but seemingly peaceful, deaths. Haraldur recounts that '[t]heir bodies were found later that morning, lying by the roadside, totally unmarked'. Various conspiracy theories circulated: a local politician was responsible for the poisonings, or these people had died because of a man-made explosion or some sort of volcanic blast.

Lake Monoun sits in a field of volcanoes and is itself a flooded volcanic crater. Hauling water samples from the bottom of the lake into their dinghy as part of their investigations, Haraldur and his student found no sign of the heating that would indicate a volcanic explosion might be the culprit. Instead, they found that the deep waters fizzed with carbon dioxide once raised to the surface. This gas had seeped out of the crater floor and dissolved in the lower layers of the lake's waters. At the higher pressures deep within the lake it remained safely locked up, held in solution by the weight of water above it. But once brought to the lower pressures at the lake's surface, the gas could no longer be contained and escaped – a process a little like volatiles escaping as a magma rises, in fact.

The team came up with the theory that a sudden overturn of the lake water, likely triggered by a mild earthquake or landslide, had raised the lake's carbon-loaded lower tiers and released a great flood of carbon dioxide. As we saw with the tales from the Cave of the Dog near Naples, this colourless, odourless gas is denser than air and so ponds into hollows and depressions. The unsuspecting villagers, carrying their wares to market in dawn's twilight, walked straight into a dip by the lake shore. There is no way that they could have detected the invisible pool of gas into which they were inadvertently plunging, and they were rendered swiftly unconscious and then suffocated by this silent killer. In a rather grim coincidence, just as Haraldur's paper on the Lake Monoun event made it to print, a similar but larger effusion of gas from the larger nearby Lake Nyos in 1986 enveloped several villages and smothered 1,746 people in their sleep.

These sinister, silent deaths from seeping volcanic gas are fortunately rare and at Lakes Monoun and Nyos their recurrence is now averted by a system of vertical tubes that vent the dissolved carbon dioxide from the lower waters at a safe rate preventing a deadly build-up in the depths. Nonetheless these volcanically carbonated lakes highlight the particular way that carbon dioxide emissions associated with volcanism can be cryptic and may not necessarily announce themselves as plumes from vents or fumaroles. Volcanoes all around the world are leaking carbon dioxide and, although sometimes a hazard as these events in Cameroon attest, these emissions have played an important role in the development of our planet's environment and our journey to being as a species.

There are many unanswered questions about the quantities and patterns that characterise global volcanic carbon emissions. Understanding these subtle but substantial emissions teaches us lessons, too, about how a planet can be both in balance in the long-term but subject to short-term and devastating swings in environmental conditions. For me, understanding this planetary-scale tension between balance and imbalance began in the here and now and with trying to understand present-day volcanic emissions of carbon dioxide. This has proven to be a task even more Herculean than I had first imagined, given the many ways in which volcanoes emit carbon and the huge challenges of measuring these emanations, especially over large areas.

One of the challenges to quantifying Earth's volcanic degassing rates is the spectacular range of ways that gas escapes the planet and, in part due to its low solubility and release from magmas deep within the Earth, volcanic carbon dioxide emissions are one of the most diverse in terms of their surface manifestations. In stark contrast to the stealthy death-clouds from Lakes Monoun and Nyos, carbon dioxide also comes out from volcanoes with a bang. Huge volcanic catastrophes, like Krakatau, Tambora or the Minoan eruption of Santorini, pumped out tens to hundreds of millions of tonnes of carbon dioxide. 'Supereruptions' like those from Taupo, Toba or Yellowstone might have blasted out thousands or even tens of thousands of millions of tonnes in a geological 'instant' – perhaps over just a few days or weeks. We saw in the previous chapter how sulphur emissions from volcanic eruptions like these can reach out to cool the world, but in contrast these rare, short-lived eruptions have negligible effects on our atmosphere's carbon budget. Large as their carbon injection might seem, compared to the roughly thousand billion tonnes of carbon dioxide in the atmosphere as a whole, they are pretty much the airborne equivalent of a drop in the ocean.

Nonetheless, unlike the sulphur gases that can cause the short-term cooling experienced after these large eruptions, once emitted, carbon dioxide tends to stick around for several hundred or maybe even more like 1,000 years. In every breath we take there is a tiny but non-zero possibility that one of the carbon dioxide molecules passing through our lungs was blown out during the massive explosion of Krakatau that rang in the ears of our predecessors and sent shockwaves reverberating around the globe. This longer atmospheric lifetime of carbon dioxide also means that its impacts are very much less dependent on the height into the atmosphere to which it is erupted than for the sulphur gases. While a short-lived released of volcanic sulphur needs to get up into the upper troposphere or stratosphere to have a global effect, the full range of styles of volcanic emissions contribute carbon that will linger in Earth's atmosphere. Although major eruptions like Tambora or Pinatubo have the greatest potential to leave an impression on the timescale of human lives and livelihoods in terms of immediate devastation or changes to weather patterns, they are also relatively rare from a geological perspective. On the timescale of carbon in our atmosphere it is more important to consider how volcanic emissions accumulate over the decades, centuries and millennia rather than focussing on these major, but short-lived, paroxysms.

We have already explored many of the array of ways in which different types of magmatic activity can release gas from the Earth. The major eruptions just discussed above certainly form part of this picture but our adventures have also shown us a rich tapestry of smaller-scale styles of emission too. This might be in booming pyrotechnics of fire fountains and lava flows on Etna or fiery bursts from Stromboli. It could be the quieter clouds persistently billowing from lava-filled vents at Masaya, Villarrica or Kīlauea. Or, softer still, gas leaks from the planet in the wafting emanations from fumaroles, as in Campi Flegrei's Solfatara or at the low summit of Nea Kameni within Santorini's bay, and, of course, the fizzing of lakes like Monoun and Nyos. Adding up all these smaller but more persistent sources spread across our planet is tricky. But, when we do our sums, they suggest that all these small emissions of carbon dioxide add up to at least as much every year as a large, but rarer, eruption like Pinatubo. As these less dramatic emissions happen year on year rather than the once a century or the even more infrequent recurrence rate of more major cataclysms, their influence on the long-term composition of the air we breathe is, in fact, collectively greater. The Earth is in constant dialogue with our atmosphere, but what it says in quiet whispers is at least as important as what it occasionally shouts.

Unremarkable as many of these carbon emissions may be in terms of the day-to-day operations of Volcanoland, they are not without their consequences. The warming effects of carbon dioxide in our atmosphere are now well publicised and understood. We read about the effect of anthropogenic carbon releases since the industrial era with increasing frequency and urgency. But this link between carbon in the atmosphere and global warming was not always a truth universally acknowledged by science. Our discourse about atmospheric carbon dioxide levels these days is focused, for good reasons, on change. That said, it is important to understand the role atmospheric carbon plays in keeping Earth's climate stable, as well as altering it. Nonetheless, our understanding of what keeps Earth regulated has in many ways been driven by trying to understand global change. Here, again, geology has played an important part. Evidence from Earth's geological past and the rock record played a key role in getting scientists thinking about how Earth's climate might vary, and what the players might be in terms of driving such fluctuations. And rather than being all about warming, it was indications of much colder periods in Earth's past that were among the first to come prominently to the fore, evidence that is literally carved into the landscape once you know where to look.

A relative used to live in a house along Under Loughrigg, a road that meanders beside the River Rothay between Ambleside and Rydal in the British Lake District. Her garden sloped steeply up into the fellside and small crags exposed between the grass and flower beds brought a sense of the wild hills above into the cultivation of a home. It was a beautiful garden and a wonderful place from which to survey the mountains, but marked into these rocky outcrops, in the scratches etching their surfaces, was the story of past landscapes too. These gouge marks or striations, which I used to trace with my fingertips while the children played among the boulders, are the trails of ancient ice sheets scouring the land. They are testimony to a long-gone time when this part of England was buried deep in ice. More fundamentally, this and other evidence for widespread ice ages led to the realisation that over longer periods than that covered by human writing, climate can change very significantly.

Many of the early ideas regarding this past glacial theory came from observations in the Alps and were championed by natural historians like the Swiss-born Louis Agassiz in the 1830s. Early observations often centred around 'erratics', boulders that are out of place and differ, sometimes very visibly, from the rocks that characterise the landscapes around them. These rocks seem to have been dumped into the terrain from elsewhere. Their size and numbers hinted at the significance and strength of the processes in operation. In the Alps, Agassiz noticed the character that glacial action imparted on the landscape. Erratics were one marker, but the ice sheets also carved wide valleys, left mounds of debris – moraines – and smoothed and scratched the bedrock with their grinding motion. He realised that these tells of glaciation engraved into places distant from present-day ice were best explained by periods of widespread colder climate rather than more localised phenomena like icebergs or flooding.

One of the early converts to the land-ice theory in Britain in the 1840s was William Buckland, then a geologist in my now hometown of Oxford. I frequently cycle past his former house in the village of Islip just to the north of the city, which was apparently crammed full of geological and biological specimens as well as exotic pets, such as a bear which roamed the village. He is also reported to have had an enthusiasm for eating his way through the animal kingdom, allegedly sampling dishes including mole, panther and crocodile. Much of Buckland's evidence in support of former glaciation in Britain came from Scotland, but he found evidence across the north of England, too, including in the Lake District. Here, as in Oxford, our paths seem to have crossed in space, albeit substantially separated by the ticking centuries of time. In his memoir reporting evidence for past glaciers in Scotland and the north of England he reports finding glacial striations on rocky outcrops in 'Dr Arnold's garden at Fox Howe [Hollow] near Ambleside', a house that stands to this day a gentle 15-minute stroll along the River Rothay from the garden where my fingers followed the ice-carved grooves nearly two centuries later.

A short hike above the Rothay's banks on a clear day gives you wonderful views from the summit of Loughrigg Fell. Imbued with our current understanding, you can read the legacy of the last ice age running through the vista. To the west the wide-based trough of the Great Langdale valley was gouged out by ancient ice, with the balcony of Little Langdale's valley hanging into its wall. Twenty to thirty thousand years ago, all would have been white, and the water that now flows freely and plentifully from the fells would have crept and strained, slowly carving at the landscape. Now the evidence seems clear, but it is a learnt reading of the terrain strung within a wider framework of Earth history. When Buckland gazed upon this landscape, the idea of widespread ice ages was still controversial, not finding broader acceptance until the mid-1870s. Buckland died in 1856 and so, never having seen this resolution, the landscape of the Lakes would have always had a different meaning to him compared to how geoscientists see it today. Everything is transformed by the filter of our attitudes, understanding and ideas, which have been far more mutable than the Lakeland hills or indeed many of the dreaming spires of Oxford over this time.

By the end of the nineteenth century, further evidence from reading the landscapes in places like the Alps had shown that there had, in fact, been several ice ages. Over the last few million years, extensive ice cover had slowly ground south and then retreated time and again. Other evidence for past climate changes came from the fossil record, ingenious techniques extracting pollen grains from sediments, and more recently, chemical analyses of ancient rocks. The acceptance of large-scale variations in global temperatures led to a need to explain what might cause them to fluctuate in this manner.

For a system as complex as a planet, especially one hosting swirling oceans, rugged continents, sweeping rivers and teeming with life, there is unlikely to be a simple solution to what drives such long-term trends in planetary change. There are many factors that affect global temperature fluctuations over a wide array of scales in time and space. The energy arriving at the Earth from the Sun is a key driver of our climate and variations in the output from our local star will have important effects. Some of the variations annotated in the geological record, like ice ages for example, broadly align with cycles in the Earth's orbital movements and how it spins on its axis, affecting how much of the Sun's energy it captures. The movement of the oceans, configuration of the land and biology and other things besides play their parts too. But another powerful control on planetary temperature is the atmosphere. In the 1820s, the French mathematician Joseph Fourier calculated that the surface of an object the size of the Earth at its distance from the Sun should be considerably colder than it is, if warmed only by the effects of incoming solar radiation. Roughly speaking, by these calculations, Earth's average temperature should be nearer 18°C below freezing instead of more like 15°C above. One of Fourier's suggestions to account for this disparity was the insulating effect of the atmosphere, making him the first to propose the importance of what we now commonly refer to as the 'greenhouse effect'.

The question of which atmospheric gases might insulate the Earth in this way, coddling in its heat, became the study of scientists like Eunice Newton Foote in the US and the Irish-born physicist John Tyndall working in London. Their careful experiments in the 1850s showed that water vapour, carbon dioxide and methane all absorb heat rays emanating from the Earth and radiate some of them back to the surface. This blocking of energy from exiting the planet was likened by Tyndall in 1862 to the holding back of a torrent: '[a]s a dam built across a river causes a local deepening of the stream, so our atmosphere, thrown as a barrier across the terrestrial rays, produces a local heightening of the temperature at the Earth's surface.' This is the basis of the science that underpins our present-day understanding of human-driven global warming.

But scientists like Thomas Chrowder Chamberlin, working at the University of Chicago, recognised that it also underpins how the Earth changes gradually over the much longer million-year timescales that characterise the ebb and flow of global temperature in the planet's past. Chamberlin saw the evolution of Earth's climate as governed by agencies. Some of these agencies, such as the chemical reaction – weathering – of silica-rich rocks by rainwater, slightly acidified via the atmospheric carbon dioxide dissolved in it, and the laying down of carbon-bearing rocks like limestone or fossil fuels, are capable of removing carbon dioxide from the atmosphere. These processes, therefore, may result in cooling and potentially widespread glaciations. Others, like volcanism, are capable of replenishing atmospheric carbon to produce warmer climates. Writing in 1899, Chamberlin's comprehension of volcanism's role in the Earth's carbon balance was limited by the then-current understanding of the origins of the gases emitted by volcanoes. It was still debated whether volcanic emanations came from deep inside the planet or were picked up from shallow sources like groundwater or the rocks of the crust. He was speculating about the potential importance of magmatism in modulating long-term climate at a time when too little was known about volcanic degassing to even hazard a guess about the scale of the emissions that might be in play. Chamberlin acknowledges that his conjectures about the importance of volcanic degassing are largely based on scientific intuition as '[t]here seem to be no means of estimating from present data even approximately the volume of gas that is given forth, but it is certainly large.'

Our understanding of the complex transfers of carbon between its different repositories within and around our planet has transformed since Chamberlin's time. Geoscientists often conceive of two interlinking carbon cycles on Earth. The short-term cycle shunts carbon between living organisms, the atmosphere, soils and the oceans. We can see this surface cycle in operation in the annual fluctuations of carbon dioxide concentrations in the atmosphere overprinting its current relentless rise. As buds burst and barren winter gives way to the green of spring for the ecosystems of the larger land masses of the Northern Hemisphere, biology draws in carbon from the atmosphere and holds this breath through the summer warmth to exhale it again as leaves fall and temperatures drop in northern autumn and winter.

Much longer-term trends in carbon's movements are governed by a deeper geological cycle. Here carbon is locked up in the sedimentary strata of the continents or on ocean floors or deeper still in the vast bulk of the mantle. Although usually sluggish to mobilise, these geological stocks of Earth's carbon far outstrip those in the shallow cycle. As a consequence, even small changes to this deeper cycle can, over time, exert great power over the long-term surface carbon budget.

One of the amazing things about this long-term carbon cycle is that it has a degree of self-regulation built into it, maintaining some sort of global balance. This is fortunate for our species and all of life on our planet. Evolution of complex life takes time, more than 3.5 billion years on Earth so far. To arrive at this point has required some bounds on how much the climate has varied. To persist and progress, at a minimum, life has needed temperatures to have been maintained within the range for at least some liquid water to be present on the surface. Several processes and runaway feedbacks that we know operate within the Earth-system could hold our planet's habitability on a knife edge. A fainter young Sun, predicted by models of stellar evolution, required more greenhouse gases to keep global temperatures above freezing, but then to keep the planet cool these gases needed to be sequestered from the atmosphere as the Sun powered up with age. Should temperatures start to rise significantly, the enhanced evaporation and ever-increasing water vapour in the atmosphere could drive a runaway warming via steam's greenhouse effect. Drawdown of the Earth's atmospheric carbon budget, driven by varying tectonics and weathering processes, could have taken the planet into a perpetual state of cold. When significant ice sheets have started to build up, their glare has reflected more sunlight back into space, thus with the potential to cool the planet ever further. At first glance, our long-term climate system seems to be in a constant state of precarious flux.

But on Earth these factors, which each *could* lead to extreme consequences, are yoked into balance by a dialogue between our atmosphere, oceans, crust and mantle mediated by weather, tectonics, biology and – importantly, too – volcanoes. Earth's long-term carbon cycle turns out to have elements of a thermostat built in. This idea has been around since the 1950s, but I first read about it in 2004 during the late stages of my PhD work. It was one of those moments that blew me away, both with the elegance of the science and the sheer luck that led to a planet like Earth evolving within the tight parameters that have allowed it to nurture us.

At the heart of the thermostat are observations made of river chemistry across the globe that show that the chemical weathering rate of silicate rocks increases with both temperature and rainfall. This weathering draws carbon dioxide out of the atmosphere and this simple observation has planetary-scale implications. Rates of evaporation and precipitation also broadly increase during warmer periods, and this invigorated weather also amplifies the consumption of atmospheric carbon by reaction with surface rocks. This means that should global temperatures rise, the rate of stripping of carbon dioxide from the atmosphere increases. Over time, this slowly diminishes the greenhouse effect, cooling the planet and counteracting the warming. Conversely, should the planet cool, less evaporation from the oceans reduces global rainfall and with the lower temperatures weathering rates decrease, both acting to lessen the removal rate of atmospheric carbon. Gradually this allows carbon dioxide from planetary outgassing across Earth's Volcanolands to accumulate in the atmosphere and warm the climate back up. Over geological timescales, atmospheric carbon dioxide both controls and is controlled by global temperatures.

While the levers of the thermostat operate on the planet's surface, it is underpinned by the faithful pumping out of carbon from Earth's volcanism. If we switched off volcanism, our planet's atmosphere would gradually deplete of carbon and the world would cool. But this will not happen any time geologically soon. Casting back into Earth's deep geological history instead of forward in time there is still much else to understand about this cycle and how it has operated. Many outstanding questions surround how the weathering rates that govern carbon dioxide drawdown have varied. But as a volcanologist, the part of the puzzle that most concerns me is understanding how much carbon dioxide volcanoes carry from the deep Earth and exhale into our atmosphere, and how this has varied over geological time.

Although we can do a much better job now than Chamberlin's intuition in 1899, totting up volcanic emissions remains a tricky task. Volcanic activity is spread out over both space and time. Volcanoes are dotted across the globe, often in remote and dangerous places, and their activity switches on and off in an irregular and unpredictable fashion. Building up a global picture of volcanic gas emissions sometimes feels a bit like guessing the image of a jigsaw from just a few pieces – and each year the picture changes too. We can fill in the gaps using statistics, but it is always informed guesswork.

For sulphur dioxide, satellite observations and other advantages that facilitate its measurement have allowed us to do a bit better, and our estimates of this gas are generally held up as the gold standard of volcanic degassing inventories. For carbon dioxide there are some specific challenges that make its volcanic flux especially difficult to estimate. The significant and well-publicised quantities of carbon dioxide in the atmosphere in general make the subtle perturbations to these concentrations from individual volcanoes harder to measure from space using satellites. The fact that carbon dioxide escapes into bubbles in the magmas especially deep within the planet means that it avoids capture in the tiny capsules of glass – melt inclusions – trapped in crystals as many of these crystals grow only after at

The author, gas sampling at the summit of Etna, 2004

least some of the carbon has already escaped the liquid rock. For the other gases, like steam and sulphur dioxide, we can use these glassy blebs to measure how much of the gas the melted rock originally carried. For carbon this is often not possible or is at least a lot more challenging.

Perhaps more troublesome still, this deep expulsion from the rocky liquid means that carbon dioxide is free to find its own way to the surface, separate from the movement of the magma. In many places around the globe, carbon dioxide steals out of magmas deep within the crust, creeping up through cracks and percolating stealthily through soils over very large areas and with few clear signs of its escape at all. These diffuse emissions are difficult to measure and estimate but critical to understand. Nowhere, I would argue, is this puzzle more complex than East Africa. Studying the rather unusual volcanism in this region gives us many key insights into how the interior of the planet works but it is also an important unknown in the global volcanic carbon budget and learning more here can also help to explain how the dance of Earth's tectonic plates might have, at times, governed its past climate.

In East Africa, the Earth's crust is slowly tearing apart. A new plate boundary is growing, unzipping a wedge of land from the rest of the continent. This process started over 20 million years ago and it has cleaved the landscape over creeping time to carve a great rift valley, running thousands of kilometres from the Red Sea and the Gulf of Aden in the north to Mozambique in the south. Volcanoes are scattered both along the valley floor and, in places off and along the rift margins. They take many shapes and forms: from majestic mountains like Mount Kilimanjaro in Tanzania to calderas like the flooded O'a (Lake Shala) in Ethiopia and many clusters of small volcanic cones in between. Many, like Ol Doinyo Lengai in Tanzania that erupts very unusual and fluid carbonate lavas, and Erta Ale in the Afar region of northern Ethiopia that hosts a spectacular lava lake, are frequently or continuously active.

If this rifting continues, one day in the distant future a new sea may flood into the heart of Africa. In the searing heat of the Danakil Depression, a plain within the Afar region at the rift's northern tip, the sun beats relentlessly down on land that is already below sea level. In places in the Afar desert faults have sliced the dusty landscape as the crust stretches so that the topography steps down like a giant staircase. Our species has a long history in this region, with some of the earliest hominin fossils, including the famous 3.2-million-year-old specimen known as 'Lucy', unearthed here. Our ancestors have long cohabited with the volcanoes and their rocky outputs that pepper the terrain. One day the successors to these volcanoes may form their own submerged mid-ocean ridge system hidden under the waves of the world's youngest sea, with a sliver of Africa from Somalia to Mozambique forming a giant island to its east. This new geography will not be delivered quickly, however, and with a likely timescale of ten million years it is sobering to think that our descendants are unlikely to be around to witness the full inundation of the land that is one of the contenders for the cradle of humankind.

The ancient origins of this rifting are likely tied to upwellings of hot mantle under Kenya and Ethiopia, something like that under Hawaii. A recent seismic study peered 2,900 kilometres into the Earth under Africa and imaged this hot plume of material. This seismic tomography returned a picture of hot mantle that looked something like an enormous billowing tree with its trunks splitting diagonally at 1,500 kilometres depth and then sprouting thinner branches vertically to stretch out under East Africa. Another branch surfaces around 3,000 kilometres southeast under the volcanically active island of Réunion in the Indian Ocean. The combined processes of hotter material upwelling from deep within Earth's rocky bulk and the stretching and thinning of the Earth's crust, reducing the pressure on the underlying mantle, contribute to it melting and to the generation of magmas and ultimately East Africa's volcanoes. It is hard to estimate how much magma is generated because only a fraction of it will find its way to the surface. In fact, the stretching of the African tectonic plate potentially makes more room to store unerupted magma than in other parts of the world. But we know that carbon dioxide gas bubbles out of magmas at tens of kilometres' depths within the Earth and the faults and fractures, again associated with the plate splitting apart, probably help to release this carbon gas even from deep stalled ponds of molten rock that might never make it to the surface. This makes it very likely that carbon dioxide is seeping out of this region on a wide and significant scale even away from obvious volcanic features. Making measurements of this diffuse flux is painstaking and time-consuming work, but also a vital piece in the puzzle of understanding the comprehensive carbon footprint of Earth's magmatism.

I started working on diffuse carbon emissions from East Africa in 2014 with a field campaign to Ethiopia to study the activity at Aluto volcano as part of a PhD project that I was involved with supervising. Aluto is one in a line of volcanoes tracing a section of East Africa's cleaving crust known as the Main Ethiopian Rift. This volcanic segment stretches roughly from Lake Chamo in the south-west to the town of Awash in the north-east. We had chosen this volcano from the many in Ethiopia in collaboration with our local colleagues and for a number of reasons. Although it has not been active in written history, there was plenty of evidence of a more tumultuous past, and satellite radar measurements showed that it was going through periods of swelling and subsidence with ground shifts of ten centimetres or more over five- to nine-month periods during the most abrupt episodes. Aluto is also close to many densely populated areas and the site of the only operational geothermal power station in Ethiopia so far. So, as well as it serving as a lens to explore our planet's deeper activity, there are strong reasons to work on understanding Aluto's behaviour to inform planning around very current risks to humans and critical infrastructure.

Aluto lies a four-hour drive almost due south of Addis Ababa, the Ethiopian capital. Addis is in the highlands bordering the rift, and at over two kilometres above sea level, you can feel the faint catch of the altitude in your breathing when you first arrive. The history of Ethiopia is deep and sometimes a little mysterious. Many believe that the Queen of Sheba was Ethiopian, and that the twentieth-century Emperor Haile Selassie was a descendant of the biblical King Solomon. Some hold that the Ark of the Covenant is housed in the Chapel of the Tablet in Axum, attended by a single guardian monk. The seventeenth-century Portuguese Jesuit missionary Jerónimo Lobo even described seeing unicorns flitting among the trees near the source of the Blue Nile.

Many of the early geological accounts of explorations of the East African Rift are tied up with European colonial aspirations in the nineteenth century. A fixation on the sources of the Nile and the pattern of accessible trade routes meant that early expeditions focused on the southern and western parts of the rift. Local written records of volcanic and earthquake activity, too, are biased away from southern Ethiopia, home to Aluto and its volcanic neighbours. Such written history is slanted towards the country's north and north-west, where the administration was usually located and where monastic libraries succeeded in preserving manuscripts through the centuries. Accounts of volcanic eruptions are further obscured by changes in place names during translations of chronicles between local dialects. The more sparsely preserved written history of its past activity makes a volcano like Aluto a very different prospect for study than those like Santorini, Vesuvius or Campi Flegrei with records going back millennia. For volcanologists, a journey to a volcano like Aluto is more like a detective mission where the challenge is to combine what previous writings there are available with local knowledge and our readings of the landscape and its rocks into the best possible picture of the system's past. Setting off from Addis for the rift that first time, two of us from Oxford, two master's students from Addis Ababa University and our driver were crammed into the four-wheel drive that bumped along the dusty roads. Despite the discomfort, the overriding feeling was of adventure and a thrill at the prospect of trying to probe some of the volcano's mysteries.

Aluto is in Oromia, one of Ethiopia's regional states. It is the homeland of the Oromo people and holds in its heart an important section of the East African Rift. Once we had negotiated the seemingly endless traffic of the sprawling capital, the road skirted south along the western flank of the rift. My heart lifted at the sight of the cone of Mount Zuqualla, which loomed like a volcanic sentry to our east as the landscape turned more rural.

This small extinct volcanic landmark is a holy site for Ethiopian Orthodox Christians and hosts a crater lake edged by forest at its summit with an ancient monastery perched on its rim. After more dusty miles, Zuqualla's companionship receded into the rear-view mirror and finally we turned east, plunging down the fall lines of the rift's boundary faults and into the basin stretched out by Earth's creeping plates.

Lakes are strung among the volcanoes along the course of the East African Rift. Aluto nestles between Lake Ziway to its north and the Lakes of Langano and Abijata to its south. Each of these lakes has their different characters. Lake Ziway is never more than nine metres deep, and from the flanks of Aluto you can see the peak of Tulu Gudo Island emerging from its waves, home to another of Ethiopia's many monasteries. The brown-tinted, mineral-rich waters of Lake Langano host tourist resorts and hot springs. The alkaline Lake Abijata feeds flamingos and a sodium carbonate production facility. The scenery and wildlife are spectacular. On a later trip, I stayed on the shores of Lake Awassa, a little further south-west in the rift. Here I would drink my morning coffee by the lapping waves, watching the incredible bird life flit around the banks in the foreground, the fishing boats at work among the reeds in the middle distance, all against the backdrop of Corbetti volcano's humped structure in the haze of the far shore. It was an intensely happy way to start a day, and it was sometimes hard to steel myself to go to work.

The lakes, volcanoes and faults have coexisted in this landscape for hundreds of thousands of years. The volcanoes have built themselves up, sometimes hollowed themselves out into calderas and, in many but not all cases, rebuilt themselves again during this period. The lakes have not been immune to change either and their levels have risen and fallen with the changing climate. Seventy thousand years ago, water levels were significantly higher and the lakes around Aluto merged into one. The waters, now confined to the lakes many kilometres to the north

Lake Awassa with Corbetti volcano in the background, taken in 2019

and south, would have surrounded Aluto on all but its eastern side. Working on Aluto's dry western flank today there is a sandy place, an alcove into the jutting volcanic rocks, which for me has the feel of certain of the beaches I have taken a dip from in the British Lake District. I have not done the proper analyses needed to confirm that this was once a shoreline, but looking out over the African landscape I can instinctively imagine throwing stones into the shallow waves as they lapped the ancient shore.

Evidence suggests that our hominin ancestors co-habited in this landscape with these shifting lakes and volcanoes. Their lives and evolution were interwoven with the changing climate and tectonics. Studies of Aluto and its neighbouring volcanoes suggest that there might have been a particular period of volcanic violence when, in a geologically brief 150,000 years, the calderas of Aluto, Gademsa to the north and O'a and Corbetti to the south all formed in large explosive eruptions. We can only speculate on the changes to the landscape and environment that these eruptions would have dispensed or how they would have been experienced by early hominin populations. Perhaps these challenges pushed their migration and development in important ways? It is hard to say.

Certainly, human eyes have long lingered on the landscape here, and our species remains a significant presence in the area. Nearly seven million people reside within a hundred kilometres of Aluto and its neighbouring volcanoes like Corbetti to the south-west. Many live in the large towns like Ziway, Awassa and Shashamane, but many others dwell in smaller towns or settlements dotted through the landscape. Around Aluto, even far from the main road, our work would often gather a crowd of onlookers with people seeming sometimes to emerge from nowhere. Explaining what we were up to – with our strange pumps, trowels and hammers – wasn't always easy, even for our Ethiopian collaborators. There are more than 80 languages spoken in Ethiopia, and our colleagues did not always speak the local Oromo language. Occasionally our activities caused suspicion, and we would be careful to move on rather than cause offence. But usually we seemed more of a source of interest and entertainment. One afternoon to the east of Aluto we unintentionally diverted a village Sunday picnic into a deep gully as we investigated the volcano's lower stratigraphy with people curious to see what we were up to and perhaps a little disappointed that it was nothing more exciting than examining the rocky outcrops and scribbling in our notebooks.

Approaching Aluto has more in common with the drive up the low ramparts of Masaya than a classic volcanic cone like Villarrica or a precipitous caldera like Santorini. Unlike at Masaya, there is no glowing lava to draw the tourists, but hints abound at the magmatism that underscores the landscape and its interactions with the environment and people it meets when it erupts. The turning from the main road to Aluto is marked by the truncated remnant of an explosive steam vent, aptly known as Adami Tullu or Cactus Hill. The shape and size of this small ring of explosive pyroclasts are formed by a collaboration between the waters of the much larger ancient lake and molten rock pushing out from beneath the lakebed. The hot magma flash boiled the shallow waters, driving phreatomagmatic (the term for eruptions powered by external water or ice, like aquifers, glaciers, lakes or the sea, interacting with magma) explosions and pulverising the rock with greater vigour than would be possible if driven solely by the magma's own steam. The cone's broken structure tells further of the power of water even after the eruption's fury was spent, with the lapping of the lake's waves eroding most of the hill away before the waters receded with the shifting climate, marooning the cone far from the present-day shoreline.

Mining trucks carrying volcanic pyroclasts and service vehicles for geothermal power projects crunch along the gravel road to the volcano as it cuts east through the white bluffs – remnants of the rain of tiny freshwater diatom algae skeletons settling on the bed of the larger ancient lake – of a muted section of the Bulbula valley. The road then snakes up the steep volcanic flank, its complex structure moulded from layers of interlocking air-fallen pumice and scoria, pyroclastic flow deposits and dense lava flows, evidence of numerous eruptions contrasting in size and style. Some of these dark lavas are formed of a jet-black volcanic glass known as obsidian and the ground around them is littered with dark chips, which glint ominously when the sun hits them, like the shells of giant beetles or something metallic. Despite their age, they look fresh enough to have been erupted yesterday. Obsidian draws its name from this region; Pliny the Elder, writing in AD 77, described it as a 'stone found by Obsidius in Ethiopia'. It is a fascinating substance. The sharp-edged, scalloped fracture patterns somehow always seem wrong to me for a natural sample, evoking the breakage of manufactured glass after an accident at home. As for human-made glass, its appearance tells of a jumble on the molecular scale, with the chaotic structure of a liquid captured as a solid. Our ancestors recognised its special properties and there is evidence that obsidian has been used to make sharp-edged tools and weapons for hundreds of thousands of years.

Over the volcano's rim, the topography still feels rough-hewn with the lobes of multiple previous eruptions building the rugged ramparts in a blocky geological collage, but in its heart you can still make out the old caldera floor, a flat plain carpeted in green and yellow and dotted with low acacia trees. When it first sank, crumbling in as large-scale eruptions of Aluto emptied out the crust beneath it about 300,000 years ago, the caldera would probably have been a sheer sided flat-bottomed ellipse stretching ten kilometres from east to west and pinched to more like five kilometres north to south. Similar topography is still preserved at Fentale volcano up in the north-east of the Main Ethiopian Rift. Echoes of this morphology remain at Aluto with some short fragments of the sheer caldera wall still distinguishable. But in the main the caldera is overgrown by newer volcanic products and cut by faults that have filled in sections of its level floor and buried its once steep walls disguising its original elliptical sensibility.

The smaller area of the flat inner landscape of Aluto that remains is nonetheless good for cultivation and is patchworked with fields and small farming settlements: clusters or single circular mud huts with grass-thatched roofs often finished with a pot at their apex, following tradition. The volcano's internal heat is also ripe for harvesting and the silver pipes of the geothermal power plant spider through the landscape glinting in dissonance with the matt finish of the savannah. Venting steam from the active wellheads puffs bright white plumes of clouds against the greens and ochres of the terrain. One of our study areas was close to one of these wellheads up in the bowl of a small valley. Its hissing roar was deafening and became exhausting before too

Aluto caldera, 2014

long. More welcome were the perpetual rainbows that danced with the sunlight off the fine spray of haze fountaining from the well and telling of the heat of the magma far below.

During that trip in 2014 I was teamed with Amdemichael Zafu (Amde), at the time a student at Addis Ababa University and one of our collaborators. We were looking to understand the patterns in the flux of carbon dioxide seeping out of the soil within a square kilometre of land close to the active geo-thermal wells. Building up this detailed picture required us to make individual flux measurements at points meshed over the scenery at 30-metre intervals. Our incremental movements were dictated by the programmed points in our handheld GPS and at every location we would stop, loosen the brown, yellow or terracotta dirt with a geological hammer and push in a short section of drainpipe so that it was semi buried in the African soil. The instrument itself was a cylindrical chamber hooked up to a sensor detecting carbon dioxide concentration. Once

mounted on its drainpipe collar, the instrument recorded how quickly the concentration of the carbon dioxide gas built up inside the chamber, allowing us to calculate the flux emanating silently from the soil.

It took us about a week to cover this square kilometre. It is rare for me to move through a landscape that slowly. I became very attuned to this small footprint of the Earth in this unfamiliar country with its steep fault scarps, steaming fumaroles and roaring plumes from geothermal wellheads. There was something intimate, too, in the tilling of the ground and the soil under my fingernails that the work entailed, reminiscent of gardening at home. We had time to notice things that we might otherwise have missed: the tiny holes in the ground, which could have been insect burrows but were actually pores through which the Earth was sweating carbon gas; the ever-varying patterns as the shards of the midday sun split between the scrubby trees caught the steam wafting from the fumaroles halfway up a steep fault scarp.

We shared our little footprint of the rift with the wildlife as well as the rocks, soil and leaking gas. Weaverbirds flitted to their intricate nests suspended from the trees, bats hung from branches tucked up in sleep, termite mounds grew from the fields, and strange insects buzzed by occasionally. But it was the baboons that really caught my attention. They would dart along ridges above us while we were making measurements and we would hear rustling and see smudges of their presence as we worked among the scrubby trees and shrubs. My fascination with their movements became something of a joke with my co-worker. Amde had grown up with baboons as commonplace, and watching my wide-eyed, childlike excitement must have been equivalent to how I would feel if someone had a similar reaction to pigeons or squirrels upon visiting the UK; he teased me a little about my nervousness when I knew they were close. His confidence was not absolute, however. One afternoon we were working our way from flux point to flux point around a

Geothermal well head, Aluto, 2014

rough, scrubby field when I started to notice the baboons gathering in the trees on the ridge of high ground to our south. More and more of them came, a line of silhouettes against the skyline until some of them got brave enough to come down to approach us and shout warnings in our direction. To my relief, Amde agreed that it was time to move on to measure elsewhere, and we watched them flood in to mingle in the field as the shadows lengthened. It seemed we had been inconsiderate enough to tramp our science through their social venue and now we had gone they could get back to their evening rituals.

But whether or not the baboons approved, each of our 1,000+ measurements ticked off in our grid was part of something much larger. Plotting up the data we could see big variations in the gas flux from the ground governed by factors like topography, the faults cutting the landscape and the permeability of the rocky surface. By combining our gridded measurements we were able to estimate the flux of gas leaking out of this square kilometre of Ethiopia that we had come to know in such unusual detail as about 60 tonnes of carbon dioxide per day. This is about the same every day as six UK households on average emit every year. Over the whole of the Aluto edifice, this scales up to several hundred tonnes per day. The big variations in the amount of gas finding its way out of the Earth just within this little square of Africa highlight the enormous challenges involved in scaling up estimates even to the footprint of a single volcano like Aluto.

We were not the only team measuring diffuse carbon emissions from the rift around this time. Another team from the UK had measured a much lower carbon flux from Longonot volcano in Kenya. A team from the US was at work further south in Tanzania's Natron basin making similar measurements. The southern end of the warm caustic waters of Lake Natron is flanked by the majestic cone of Ol Doinyo Lengai, perhaps one of the world's most unusual volcanoes. As briefly mentioned above, some strange chemistry deep within the

Earth has created magmas that are rich in rare types of sodium and potassium carbonates rather than silicate, and gush like water during Ol Doinyo Lengai's frequent eruptions. Perhaps unsurprisingly given this evidence of unusual carbon plumbing in this part of the rift, the US team measured high fluxes of carbon seeping out of an extended area here presided over by Lake Natron's glorious, but endangered, indigenous flamingos.

Although the stories from lakes Manoun and Nyos are a warning, on the whole, the hazards to people and animals from this quiet seeping of carbon dioxide from the ground is limited, as long as poorly ventilated hollows are avoided in areas of strong degassing. The impacts on the local environment are minimal, too, with little visible damage other than the areas within about an arm's length of the fumaroles themselves. Nonetheless, to understand our planet and how its internal machinations have driven its surface environment estimating the scale of the carbon dioxide emissions leaking out over the full length of the whole East African Rift Valley is important.

Quantifying this flux is not straightforward. As described above, the time-consuming nature of the measurements mean that we only have information from a tiny fraction of the vast expanse of the rift's footprint. Working out the best way to tile these relatively small and geographically fragmented measurements together to produce the best rift-scale estimate of carbon emissions over the thousands of kilometres that it stretches is very challenging. There is still a great deal of work to be done but our best guess at the moment is that something like 30 million tonnes of carbon dioxide gas is leaking silently out from the deep Earth in East Africa each year. There are big error bars on this: some estimates are three to four times higher and some are ten times lower. There is speculation that rifting like that seen in East Africa has the potential to mobilise usually inaccessible reservoirs of carbon from the brittle upper part of Earth's mantle under Africa. If the higher estimates of carbon dioxide flux turn out to be true, this would support this extraordinary process. We have made progress but there are many mysteries still to unravel by toiling through the rift's stunning landscape.

East Africa is both an important global region and a key place to study continental rifting in action. But volcanoes in all of Earth's tectonic settings are emitting carbon. Studying the different behaviour of carbon associated with these different types of volcanism in the present day again helps us to understand these processes and their long-term atmospheric impacts. Magmatism at mid-ocean ridges can bind seawater carbon into the new oceanic plate that it creates, as well as emitting it from the billowing submarine lavas. This capturing of carbon limits its net flux from ridge volcanism. In fact, mid-ocean ridges could be carbon sinks rather than sources in scenarios where more carbon is bound in than outgassed. Hotspot volcanoes like those of Hawaii vent carbon transported from the mantle. Subduction-zone volcanoes are chimneys for the mantle, too, but their carbon dioxide emissions may be augmented by carbon that plunged down into the Earth's depths with the sinking tectonic plate that drives the magmatism. This is likely especially where carbonate rocks like limestone build the crust of the subducting plate, as is the case in Central America under volcanoes like Masaya. In other places, like Italy's Etna, Greece's Santorini or Mexico's Popocatépetl, the carbon-rich gases bear the signature of crustal limestone heated and decomposed into the magmas at some stage during their journey from mantle source to eruption.

Work across the planet by volcanologists and geochemists means that we can do an ever-better job at assessing Earth's total volcanic carbon degassing flux. Diffuse emissions have been mapped in many regions by similar techniques to our studies in Ethiopia. In other locations, scientists measure fumarolic emanations (as we did at Solfatara in Italy in Chapter 4), hot springs like in the shallows around Santorini's Kameni islands (as described in Chapter 4), and bubbling mud pots in places like the US's Yellowstone or New Zealand's Rotorua. In strong and visible plumes like at Masaya we can use spectrometers to look for the infrared signature of carbon dioxide, and more recently, flying drones into the plumes from more dangerous or inaccessible volcanoes has opened up new opportunities to count up the planet's carbon-rich breath. As for East Africa, our calculations come with wide error bars, but our current best estimate puts the total at, at most, about half a billion tonnes of carbon dioxide per year. This number is important for many reasons. From everything that we can deduce from the geological record, it does not appear that we live in particularly extraordinary times in terms of rates and magnitudes of volcanic activity. This present-day flux gives us a sense of scale in terms of the influence of volcanism on the carbon balance of our atmosphere during relatively unremarkable – geologically speaking – periods of planetary magmatism.

Our understanding of the underlying processes that might subtly change these fluxes allows us to hypothesise about how volcanic carbon outgassing might have meandered around this average over Earth's history, and what consequences these variations might have driven. For example, every measurement that Amde and I made scrambling around through the dust and the scrub at Aluto during those February days in 2014 – and those of other measurement campaigns like ours – while unique in Earth's history also have far wider implications. In the tectonic dance of Earth's plates, continents have come together but also torn apart. Evidence tells of great supercontinents in our planet's past, when most of the land was clumped together leaving a vast blue mass of ocean at its antipode. The most recent of these was Pangaea, which ripped apart in stages during a period of about 100 million years starting in the Early Jurassic/Late Triassic Period (approximately 200 million years ago). During these periods of continental break-up, the physical length of the rifting, slicing up the land and carving new oceans, would have been far greater than the total of the world's rifts today. As in East Africa in the present, these ancient rifts would have leaked gas and scientists hypothesise that the elevated emissions of deep-sourced carbon dioxide during these periods of increased rifting in Earth's past might have driven climate warming.

There are many challenges in reconstructing the arrangements of continents and oceans many hundreds of millions of years ago. But to explore the connection between rift degassing and atmospheric carbon levels through geological time, we also need to know how to make a best guess at the amount of carbon dioxide that typically comes out per kilometre of a splitting continent. Tilling our drainpipe collars into the dry Ethiopian soil to determine a gas flux, Amde and I were playing our small part in building up a picture, data point by data point, that helps to unlock the carbon footprint of supercontinental demise. Other tectonic changes will also alter the amount of carbon dioxide that the planet degases over geological time. These include variations in the total length of global subduction zones, the cargo of carbon-rich rocks like carbonates being subducted that then enrich the resulting magmas, and fluctuations in mantle hotspot activity.

We again get a sense of the scale of these effects by understanding volcanic processes today, and once more learning from the present helps us to explore how the Earth's climate might in part – although, of course, not exclusively – dance to the same beat as its drifting tectonic plates over geological time.

Our total present-day volcanic flux of half a billion tonnes of carbon dioxide per year also gives us a yardstick against which to measure other things that might change this balance more dramatically. In Chapter 8, we will adventure through the evidence of periods of volcanism in our planet's past that far surpass anything in recent times, leaving features known as large igneous provinces behind on Earth's surface. But there are

Amde and the author gas sampling, at Aluto in 2014

other non-volcanic ways that the world's carbon cycle can get out of kilter. Large asteroid impacts, like the huge 10-kilometre-diameter chunk that created the massive Chicxulub crater in Mexico at the close of the age of the dinosaurs (at the end of the Cretaceous Period, about 66 million years ago), can vaporise swathes of crustal marls (carbonate-rich muds), limestones and other carbon-rich rocks, kicking up global temperatures. This might mobilise hundreds or possibly thousands of billion tonnes of carbon dioxide into the atmosphere, contributing to the environmental catastrophes that may follow. More recently, our total global volcanic flux gives us a benchmark against which to judge our own post-industrial release rate of fossil carbon.

Despite the wide error bars in our estimates of the global rate of volcanic carbon degassing, what we can know is that these natural emissions pale into insignificance compared to what humans produce. In 2019, human fossil-fuel burning released over 35 billion tonnes of carbon dioxide into our atmosphere. This is 70 times more than even our most generous current estimates of global magmatic carbon degassing. In 2022, the aviation industry alone emitted 800 million tonnes of carbon dioxide, eclipsing estimates of that from our planet's background tectonism before even considering other sectors of human industry. We cannot look to Earth's volcanism today to reassure ourselves that our rate of carbon emission might not be too much of a change in terms of our planet's natural cycles. Powerful as the forces of tectonics that daily drive the slow creep of plate movement and volcanic activity across the globe are, the human race has currently surpassed them in terms of its carbon dioxide flux to the atmosphere. It is apposite to reflect upon the level of responsibility that should appropriately come with the level of power attained by our species that, by this carbon metric, overwhelms all Earth's volcanoes.

Over geological timescales, Earth has recovered from past carbon crises like that caused by the large asteroid impact marking the end of the Cretaceous Period, the periods of supercontinent break-up or the huge volcanic outpourings we shall explore in Chapter 8. This ability to recover is, we believe, importantly, largely mediated by the slow weathering of rock by rain drawing down atmospheric carbon dioxide in the cycle described earlier in this chapter. I have no reason to believe that Earth will not slowly recuperate from wherever our addiction to carbon takes us. This will not, however, be on a timescale that is at all useful for us in terms of mitigating the consequences of our collective actions. It is nonetheless something I take solace from when I am anxious about the state of the world – even if it doesn't alleviate a strong wish that we might do better at reining in our emissions sooner to avert the changes and suffering that continuing will likely cause.

I once caught a day flight back home from Ethiopia and was pleased to get a window seat. I was tired after the trip and so let my head loll against the window and watched the verdant green of the deeply gullied Ethiopian highlands brown into the tans and ochres of the desert as we struck north for Europe. Our flight path took us directly over Khartoum, where the Blue and the White Nile join to form the weaving trunk of this mystic river which cuts onwards towards Egypt through the barren terrain. Even from cruising altitude the desert scenery and the stripe of this mighty river were mesmerising. I had come straight from the rift and from the little kilometre square of degassing Earth on Aluto of my intimate acquaintance. Weary as I was, there was something satisfying in imagining the possibility that atoms of carbon that I had sensed invisibly wafting out from the soils and fumaroles of the rift might one day be captured and carried in the coursing river below and sequestered back into the Earth.

The irony of pondering this balance accompanied by the roar of the aircraft engines fuelled by burning fossil carbon was not lost on me. My wonder at our world and its lucky state of balance was tinged with heaviness in the knowledge that Earth's ancient thermostat cannot help us escape the responsibility of dealing with our own power to change the climate. We are fortunate to have a world that is balanced in the way that it is, and Earth's present-day volcanoes guide us about its bounds of short-term carbon tolerance. We do not need to look far to see the outcomes of different pathways of planetary evolution. Earth is not the only planetary body to have volcanoes, and they have played their parts in environmental evolution on different worlds too. Casting our eyes beyond the thin blue line of our atmosphere brings home truths about how different things could have been and the precious path that Earth charted to make human life possible. Volcanoland, it turns out, stretches into the stars, and while we cannot visit we must cast our imaginations out there to understand and appreciate all the more clearly what we have here on Earth. It is out into the heavens that we shall journey next.

Part Three

The reach of Volcanoland through outer space and deep time

Chapter 7

Home Truths from the Heavens

What is the nature of volcanism on other worlds?

Volcanic landscapes can seem very alien. The word 'moonscape' often springs to our lips when describing trips to volcanoes and there is often a strong sense of other-worldliness that pervades these places. Fresh volcanic landscapes are primitive and their inorganic forms and hues can make us feel like intruders into a foreign realm. Biology is such a defining feature of our home world and its apparent absence or covert presence in a terrain recently coated or scoured by volcanic dust or fumigated by acrid volcanic gas might be one reason why areas of active volcanism can feel so extra-terrestrial. But you do not have to look far to see evidence of the importance of magmatism on other celestial bodies. If you gaze up at the full Moon on a clear night, the dark patches, or maria (mare in the singular), which give our cosmic companion its familiar countenance, are not the watery seas imagined by early astronomers but rather great ancient outpourings of basalt that occurred several billion years ago. The lunar maria cover about 15 per cent of the Moon's surface and are mainly on the side tidally locked to face Earth. Given the darker, less reflective nature of the mare basalts, compared to the mineral-glint of the feldspar-rich highlands, this means that the so-called dark side of the Moon would, in fact, be brighter than our familiar lunar vista if only we could see it.

To look to the heavens in search of meaning seems a profoundly human thing to do. Archaeology is peppered with evidence of the very earliest human societies venerating the Sun, Moon and stars, and often, like volcanoes, deifying them. To cast beyond our atmosphere and find that volcanism transcends the borders of our planet is not a surprise for us given our scientific understanding of the Earth. The rocky planets and their moons were made from the same basic ingredients as Earth and share molten origins in the tumult of the early solar system. This similarity to our home planet in terms of their genesis and make-up means that magmatism is very possible, if not likely, at least during some stages of their long planetary lifetimes given the abundant volcanoes we see on Earth. We readily see evidence of past volcanism. Most of the Moon's volcanism is long dead, but to gaze at the dark maria framed in its full circle is to see, in an instant, more lava with the naked eye than even the best-travelled volcanologists will manage on Earth in a lifetime: a nugget of contemplation an American planetary scientist friend handed me many years ago as we shared reflections under the full Moon in a clear autumn sky one evening in Oxford. I now think of it each time I catch the Moon's bright orb wherever I might find myself around the planet.

There is, of course, a limit to what we can discern with our vision alone. Technology allows us to probe further, though, and we are discovering ever more about the other astronomical objects of our solar system and beyond. We have learnt much about volcanism and its importance here on Earth, but as our sightlines stretch ever further, it has never been timelier to ask questions about what connections we can make between our own geology and that of alien worlds. This exploration started with planetary bodies in our own solar system but, more recently, extra-solar planets (also known as exoplanets) orbiting stars other than our local sun are a fresh frontier and new measurements and discoveries ask increasingly expansive questions about how far Volcanoland might extend into the heavens. In this brave new universe, what then can our understanding of volcanism on Earth tell us about its characteristics and effects on other worlds in our solar system and beyond? But, as with the awe we feel under the great expanse of a starry night sky, looking out is also an act of introspection. As well as using lessons from our own planet's geology to understand new planets, there are also many lessons to be gleaned from the environments of these other worlds about the precarious balance, in part mediated by volcanoes, that we now host here on Earth.

Our study of volcanism on the Moon goes back centuries. In the early seventeenth century, Galileo Galilei, the polymath from Pisa, picked out craters, mountains and valleys on the lunar surface using the newly invented telescope. Lunar craters were thought for a long time to be volcanic in origin. Better telescopes brought more detailed observations. These improved images and greater understanding of what happens when asteroids or meteorites impact a planet's surface, volcanic pits and, rather darkly, nuclear bomb craters here on Earth, eventually led to consensus that the origins of these lunar pockmarks were via external bombardment of the Moon's surface rather than volcanism. Meandrous channels (known as 'sinuous rilles') associated with the maria on the lunar surface have been documented since the late eighteenth century and, although still enigmatic, are often interpreted as evidence of past lava flows.

By the time Neil Armstrong took his 'small step' onto the lunar surface in 1969, there were already many ideas about the Moon's place in Volcanoland but getting human boots and eyes onto our satellite's surface was transformative in our understanding of its volcanism as well as being among our greatest feats of space exploration.

Contemplating the Moon in the night sky, it still amazes me to imagine that 12 different people on six separate missions successfully landed and surveyed its distant surface in person. For all the incredible pictures sent back from landers on more distant places, it is still the only other world of which human beings have first-hand knowledge. Although only one formally qualified scientist (Harrison 'Jack' Schmitt on *Apollo 17*) has so far walked on the Moon, all of the *Apollo* astronauts underwent training in field geology in Moon-analogue environments on Earth like Iceland and Hawaii. The Moon, after all, remains in many ways the ultimate geology field trip to date especially for those of us with an igneous bent. The desire for 'just one more rock' is a trait that touches many geologists when sampling in the field and the story of Dave Scott on *Apollo 15* suggests that the astronauts certainly took on some of this rock-lust psychology. Scott noticed an interesting-looking rock on the surface while driving the Lunar Roving Vehicle and stopped to collect it. This kilo of rock is officially labelled Lunar Sample 15016, but is better known as the 'Seatbelt Basalt' because Scott claimed to be just stopping to refasten his seatbelt while he bagged it, assuming that mission control would not otherwise grant permission for this delay. Once imbued with the love of rocks, an interesting find is a wrench to leave behind, whatever part of the universe you find yourself in.

During the six successful lunar-landing missions, NASA brought about 380 kilograms of lunar rocks, core samples, pebbles, sand and dust back to Earth. Although we also have meteorite fragments identified as from some other planetary bodies, these 2,200 separate Moon samples, along with small quantities of rock from three Soviet robotic spacecraft in the 1970s and more recently the Chinese robotic *Chang'e* 5 in 2020, form the majority of the extra-terrestrial rocks our species so far possesses where we know the details of their exact original location on another world. These precious rocks tell us many things about Earth's orbiting companion. Dating them by measuring the radioactive clocks caught up within them defines a lunar geological timescale. The bright lunar highlands are amazing in their antiquity, with rocks dating back 4.3 billion years and possibly beyond. The crystals that form the highlands and reflect the Sun's light back at us so brightly in the night sky likely once bobbed to the surface from a near global magma ocean where they cooled to form this lustrous crust.

The dark lunar maria poured forth later, mainly between 3.1 and 3.9 billion years ago, in great deluges of basalts largely flooding older impact basins. Although some recent studies have suggested that there are small basalt flows on the Moon less than 500 million years old, it is more commonly accepted that most mare volcanism had ceased by one billion years ago. The broad features of the Moon's visage visible from Earth have likely changed little since life first evolved eyes to gaze upon it. The reasons for this very different thermal history compared to our geologically vibrant home are locked up in the Moon's more diminutive mass (a hundredth of that of Earth) and size (a third of Earth's diameter). This smaller size binds in less internal heat energy (both primordial and from radioactivity) and gives it a higher surface-area-to-volume ratio meaning that, broadly speaking, it loses this heat more efficiently. So, the Moon's insides have cooled faster than our world, and volcanism has ceased. This degree of internal cooling will eventually happen to Earth, too, but not for billions of years. We have far more pressing things to worry about, but it is another moment of sobering perspective when viewing the Moon's bright disc in the sky: we are in some ways looking into Earth's geologically more torpid future, as well as the lunar past in the light reflected back from its ancient and now tectonically static moonscape.

Observations of the Moon and its rocks allow us to test our theories of planetary and atmosphere formation and volcanic processes on a system with similarities to but also significant differences from Earth. Determining the quantities of potentially volatile gases tied up in the Moon's rocks is crucial to all these projects, and whether or not the Moon holds significant water in its rocky interior and on its surface (as the maria-scale oceans imagined by our predecessors or more recently in possible measurements of lunar ice deposits) has been debated for centuries.

There are still many mysteries about the details of the Moon's origins, but for virtually all plausible theories, such as its formation during a giant and violent impact between the proto-Earth and another roughly Mars-sized protoplanet, it seems likely that water and other volatiles present in the original Moon-forming material would have been boiled off during the heat of its creation. Analysis of the minerals and the tiny melt inclusions trapped in crystals in the *Apollo* rocks (similar to the measurements I described applied to rocks from volcanoes like Kīlauea in Hawaii) allowed scientists to test these theories. Early measurements of low dissolved-water contents in the rocky lunar glasses supported such predictions. But more recent measurements using more advanced analytical techniques show higher volatile contents, very similar to primitive terrestrial mid-ocean ridge basalts (pushed out where Earth's tectonic plates move apart deep beneath the oceans) in some lunar samples, suggesting a more complex picture and opening new questions about lunar evolution.

The lack of clouds, twilights or other atmospheric phenomena had convinced Moon observers from Galileo onwards that it was not swathed in anything like Earth's gassy envelope. Although for most practical purposes, the Moon can be considered to inhabit space's vacuum, measurements on the *Apollo* missions suggested a thin vestige of an atmosphere with a pressure about a million billion times lower than Earth. Whether it had a significant atmosphere in its past depends on whether the venting of gas from its interior ever outpaced the loss of gas atoms and molecules to space. This rate of 'atmospheric escape' is higher for the Moon than for Earth, given the weaker pull of its gravity, and helped further by the stream of charged particles flowing from our Sun – the solar wind – an effect mitigated on Earth by our stronger magnetic field.

Estimates of the concentrations of gas dissolved in the Moon's rocks returned during the *Apollo* missions allow us to calculate the degassing budget of lunar volcanism as it has waxed and waned over time. There are still many unknowns, but calculations show that there might have been a more substantial atmosphere, albeit still insignificant compared to Earth's, in the Moon's past – pumped out during the height of lunar maria eruptions. This lunar 'air' would have been ephemeral and sputtered away into space in a geologically short length of time. As we saw in the last chapter, volcanism's enduring dialogue with Earth's atmosphere is a key contributor to its long-term composition and balance. The Moon highlights active magmatism's hand in the very existence of an atmosphere.

The amount of water in or on the Moon is important, too, in the understanding of the style of its volcanic eruptions. As discussed in earlier chapters, by comparing the characteristics of volcanic activity across Earth's volcanoes, we have formulated models to explain the differences in characteristics like explosivity, lava run-out lengths and volcano shape. These models are frameworks for our understanding that allow us to make predictions about new volcanic events and are based on factors like magma chemistry, speed of ascent in the Earth's crust and initial dissolved water (or other volatile) content. Earth's tectonic cycle, atmosphere and the strength of its gravity have formed an inevitable backdrop against which this understanding of volcanism has grown and sits.

The Moon has no plate tectonics, negligible atmosphere or hydrosphere and greatly reduced gravity. Without subduction dragging oceanic water into its interior, magmas will always contain less dissolved water than those driving the fearsome eruptions from subduction-zone volcanoes like Tambora or Krakatau. Low dissolved water and other gases in its magmas might lead us to predict that all volcanism on the Moon would ooze out with little explosive fizz. But, in fact, there is evidence for at least mildly explosive eruptions on our celestial neighbour. Tiny spheres of volcanic glass returned in the *Apollo* soil samples tell of ancient fire fountains. Mapping of the lunar surface suggests over 100 locations where such explosively born volcanic pyroclasts mingle with the dusty soil, sometimes possibly scattered significantly beyond their vents. Low as dissolved gases might be in lunar magmas, their eruption into a near vacuum sucks them out of the magma to form bubbles more effectively than on Earth. With no atmospheric pressure to brake their swelling, once these bubbles have fragmented the magma, their expansion is essentially unbounded and hence fast, maximising the explosivity even from their low volatile load. The lower gravity on the Moon enhances their range and dispersion. If we had been able to cast our eyes to the Moon looming large in the early-Earth night sky billions of years ago, its milky crescent might have framed patches of fiery red, radiating dark smudges of fountaining material.

Over the centuries we have developed many techniques to study the plentiful supply of volcanic rocks on Earth. More recently, probing the precious samples returned from the Moon using these same technologies yield rare messages drawn directly from another world. Our species has been mesmerised by the sight of the Moon in the night sky throughout our history. Every new insight that science gives us today makes me gaze in ever deeper wonder. But despite its shared history with the Earth, clear evidence of past volcanism and the amount we have learnt from observing and even visiting it, the Moon's small size and mass mean that it is never going to be the best analogue for Earth and the processes that have driven its evolution. To understand more about our planet and the role of volcanism in its development we must cast our eyes further, first to our nearest planetary neighbours – Mars and Venus – and then further still, into the outer solar system and beyond.

No human has ever set foot on any other planet. The clear candidate for this next step into the universe is the bright pin prick of slightly red-tinted light the ancient Egyptians called simply 'the Red One' and we call Mars. Nowhere on Earth is quite like Mars, but to prepare for past, present and future Mars exploration via probes, rovers and maybe one day human missions, terrestrial Mars-analogue environments are often used to test sampling techniques and equipment, especially in the search for any relics or hints of biology that our red neighbour might hide. Mars's surface is dry, cold, bathed in intense ultraviolet radiation and its soils and dust host pervasive salty chemicals such as perchlorates, chlorates and sulphates. The plateau of the Atacama Desert, sandwiched between the Pacific and the high Andes, shares many of these attributes. With an average rainfall of only 15 millimetres per year, it is one of the driest places on Earth. 'When it rains here, people die' a colleague from the University of Antofagasta explained many years ago when we passed through on our way into the Atacama. These words echoed in my head in 2015, when 14 years' worth of rain fell on the region in a day, turning parched soil into thick, muddy torrents which killed at least 26 people and swept away 2,000 homes. It is this aridity, the cold of its altitude, the intense sunshine slicing through the thin largely cloud-free atmosphere and the unusual chemistry of the region's salt pans that make it such a Mars-like pocket of Earth and one of the global areas often used as an analogue for the red planet.

I visited this region of Chile in 2003 as part of the same field campaign that took me down to the wet Chilean Lake District described in Chapter 5. Although this particular expedition had no direct relevance to Mars research, this trip still stands out in my memory as evoking an enduring feeling of being out of this world. Maybe it was because I was aware of the interest in the area by NASA and other scientists to inform the search for extra-terrestrial life. Maybe it was bonding with an astronomy PhD student about life in the Atacama before we left – the altitude and clear skies making it a prime location to site the huge communal telescopes built to interrogate the heavens over the past few decades. Maybe, too, it was the fact one of my colleagues on that trip was a former US Air Force pilot now training in geology, whose dream was to be among the first NASA crew to Mars. These things said, I do not think that the Atacama scenery needs anyone or anything else to speak on its behalf in terms of its extraordinarily vivid beauty or its power to imprint a sense of something alien and unyielding to those of us unaccustomed to its landscape.

The colours, sensations and scale of the Atacama are like few other places I have been. It is a land of rusty and ochre rocks, and mountaintops dusted with snow or strange salts under the gaze of a cool but unforgiving sun. In some places patchy vegetation clings on, in other places it is entirely absent. Herds of llamas and families of ostrich-like rhea roam and rabbit-like rodents called viscachas peep out from between the boulders of lava flows. Salt flats stretch across parts of the landscape. The largest – the Salar de Atacama – carves about 3,000 square kilometres into rough dirty-white topography by evaporation from pools of briny water. Away from the larger salars, smaller salt-rimmed brackish lakes mirror the sky between the mountains, reflecting its blue with strange tints of green and turquoise added by their mineral waters. Flamingos are sometimes visible dipping into their mirage-like shallows. On one excursion we bumped along a dirt road extra slowly, unintentionally driving a flapping flamingo in front of us until it finally found lift-off into flight.

While the Atacama's beauty is uplifting, it also leaves you in little doubt that it is a relentless and harsh environment. The air is dry and thin. It is easy to dehydrate, and to begin with, even minimal exertion can leave you panting. Isabel Allende expounds a notion that it is this landscape that carved the seriousness that she and others perceive in the Chilean psyche. This cultural countenance, in Allende's words, a bequest from the 'exhausted Spanish conquistadors, who arrived half dead with hunger and thirst' having slogged 'across the Atacama Desert beneath a sun like burning lava'. It is important to remember though that, despite the unforgiving conditions, there were people with a long history of living in this area far before the conquistadors' arrival. The Chinchorro people, for example, left mummified human remains dated by archaelogists to over 6,000 years ago and the indigenous Atacameños are still very present in the region with a culture tracing back at least 1,500 years.

Although Allende deems that the reward for the conquistadors' struggle through the desert might have seemed 'scarcely worth the trouble' with Chile not yielding gold and silver like other regions of the South American continent, the bounty of this harsh terrain turned out to lie in different mineral riches. Vast copper mines now carve into the region, and one of the world's largest lithium reserves hides in the brines under the salty hummocks of the Salar de Atacama. As global demand for such resources accelerates, so too does the threat to this incredible environment. There are some very earthly dilemmas between global human needs and those of the local people and environment playing out in this other-wordly Martian analogue.

*

Although not central to the reasons for considering the Atacama a Mars analogue, volcanoes are another thing that it has in common with the red planet. In the area around San Pedro where we were working, volcanism has been a powerful force in defining the character of the landscape. To the east of the Salar de Atacama, the great ridge of the Andes is punctuated with volcanic peaks built of magmas seeded from the subduction of the oceanic Nazca plate beneath South America more than 350 kilometres to the west beneath the Pacific's waves. Four million years ago, a massive eruption tore through this landscape, leaving immense sheets of dense welded pyroclastic flow deposits like those on Santorini but on a much larger scale. This eruption from the La Pacana caldera is one of the five largest single eruption events that we know about in the geological record. There have been no such large-scale eruptions in recent times in the area but it is far from volcanically quiet. In 2003 we were there to study Lascar volcano, one of the most active in the region. Its eruption in 1993 sent debris up to an altitude of over 20 kilometres, high enough for light ash fall to be reported in Buenos Aires more than 1,000 kilometres to its south-east. Pyroclastic flows travelled 7.5 kilometres to its north-west, requiring the people of the small settlement of Talabre to relocate their village to higher ground to minimise the threat from future eruptions. Fortunately, at the time of writing, no recent eruptions from Lascar have matched the scale of that in 1993, but its summit crater still hosts an active lava dome with intermittent smaller explosions and continuous degassing. It was these high-altitude emissions of gas and fine particles from its summit, rising 5,592 metres above sea level, that were the target of my field studies.

Our base camp (at 3,580 metres altitude) while studying Lascar was Talabre's small schoolhouse, which was empty during the summer break. The villagers (who numbered 64, according to Talabre's sign) largely left us to ourselves and although our breeze-block abode decorated with school posters and children's art bore negligible resemblance to any dwelling we might imagine on Mars, the small size of our team and Talabre and our location on the edge of the mountainous wilderness of the Andes gave me some very small sense of the isolation that Mars-analogue crew experiments seek to replicate. This cluster of stone dwellings with tin roofs was quite literally the end of the road, or at least the tarmacked road. It was often windswept and dusty, with Lascar looming to its east – an ever-threatening presence. But it was a magic place too. Here you could creep out into the chilly air of the Andean dawn and watch the sun rise, and pick out in pink the persistent fumes wafting from Lascar's summit and casting strange horizontal crepuscular rays westwards around the silhouette of the broken mountain. Here occasionally llamas roamed the streets, the daytime colours were raw and vivid and the night skies seemed alive with starlight.

Life in the schoolhouse was a step up from camping in the chilly desert night, but not without its challenges. The building

Talabre cemetery with Lascar in background, taken in 2003

had one dripping shower drawing from a water tank on the roof filled from a stream fed by meltwater from Lascar and its neighbouring volcano. There was a strict rota in terms of getting the first shower after a day in the desert dust, as only the water that had sat in the tank all day had had the biting cold removed from it. Everyone apart from the first taker risked a rapid wash in a painfully icy spray. Charging equipment had to be done with urgency, as the village only had power for a few hours each evening when the central generator was switched on after dark.

Given their proximity to Lascar, the people of Talabre were well used to volcanologists basing out of their village and, as I mentioned, for the most part we drew little attention. There was far more excitement during the fortnight of our stay about the visiting missionaries who ran through the full year of Catholic masses over the course of a week in the small church hall close to the school. Catholicism was brought to Chile with the conquistadors in the sixteenth century but Allende relates the broader spiritual compulsion of the Chileans to the landscape rising 'from the earth itself: a people who live amid mountains logically turn their eyes to the heavens'. I imagine that the power of this impulse is likely amplified when you live high up in the sky next to a capricious volcano.

That we are able to cite the Atacama as a good Earth analogue for the Martian environment is in itself a significant achievement of human endeavour. The ability to see beyond what is visible to the naked eye and know that even if we have still to find clear evidence of other life in the universe, Earth's volcanoes at least are not alone, is no meagre feat. Our understanding of volcanoes on other planetary bodies beyond our Moon comes almost entirely from instrumented probes that we have sent to slingshot through the solar system since the 1960s.

When the *Mariner 9* spacecraft first went into orbit around Mars in 1971, the whole planet happened to be engulfed in a violent dust storm. The tops of four mighty volcanoes were the only visible features emerging from the turmoil of the swirling clouds. The greatest of these peaks, towering 25 kilometres above its surroundings, had already been picked out as a bright spot on the Martian surface through telescopes from Earth as far back as the nineteenth century. It is now known as Olympus Mons and it is not only the largest volcano so far discovered in our solar system but among the largest mountains. It blisters up on the boundary between Mars's lower northern plains and southern highlands. Its shape is reminiscent of the great shield volcanoes of Hawaii and its summit crater and caldera complex looks not unlike Masaya from above. But it is on an entirely different scale. Its huge bulk is aproned by an enigmatic escarpment of up to roughly ten kilometres, making it almost look as if this giant complex structure landed intact on Mars's surface, stamping out a footprint the size of the US state of Arizona. Its scale and shape make it nothing like any volcano we have mapped on Earth and its huge volume can be accounted for by the slower erosion rates and the lack of plate movement on Mars. This means that a mantle hotspot, like the one feeding Kīlauea today, has grown just this single monster edifice (rather than a chain of volcanoes like the Hawaiian islands) pushing out lavas at a similar rate to those on Earth but at the same location over a billion years of Mars's history.

Volcanism has played a significant role in the evolution of Mars. Since the earliest probes beamed back pictures in the 1970s, it has been apparent that volcanic features cover large portions of Mars's surface. As well as huge volcanoes like Olympus Mons, lava flows and plains swathe vast regions. There are indications, too, of the friable deposits from explosive volcanism but so far still significantly less than the evidence for effusive eruptions like lava flows. As on the Moon, the lower atmospheric pressure on Mars (about one per cent that of Earth at the surface in the present day, although it might have been much greater in the past) ought to aid the formation

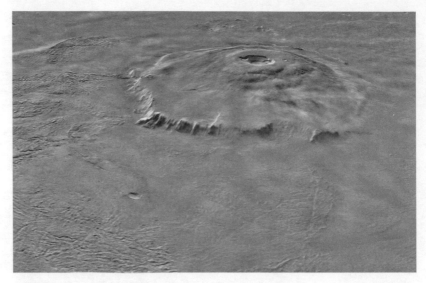

Orbital view of the Olympus Mons volcano on Mars

and expansion of bubbles in an ascending magma, increasing explosivity. On Mars, unlike the Moon, there is abundant evidence that, back in its ancient history, liquid water flowed and carved the planet's surface. NASA's *Curiosity* rover has explored in some detail the geological remnants of a dried-out freshwater lake, sending back stunning images of smooth pebbles rounded as they rolled for kilometres downstream in a river that was ankle- to -hip-deep. As well as opening the intriguing possibility of habitable environments in Mars's deeper history, this reservoir of surface water might also offer extra fuel for explosive volcanism in Mars's past. Like in the example of the cactus-adorned tuff cone left dried out by the receding water of the shallow lakes in Ethiopia's rift valley described in Chapter 6, Martian waters (or, more recently, subsurface ice) could have flash boiled when touched by magma pulverising the hot rock into a rain of fragments.

But, as Earth-like as some of the landscapes imaged and beamed back to us by ever more sophisticated rovers are, there are no lakes or rivers on Mars today. Studies suggest that significant quantities of Mars's water was lost to space or is locked up in minerals within its solid bulk. With an average surface temperature of about -60°C, on the present-day planet the surface water that exists is mostly trapped in the polar ice caps or frozen under the surface. Nor do we see any evidence for significant active Martian volcanism today, although it is possible that there might be some very low-level activity going on. Instead, evidence suggests that rates of planetary volcanism peaked over three billion years ago and have since been in decline. Huge as Olympus Mons is, its major activity looks to have ceased over 2.5 billion years ago. Although larger than the Moon, Mars is still smaller than Earth with only a ninth of its mass. For Mars, as for the Moon, this earlier slowdown of volcanism compared to our home is explicable given its smaller reservoir of primordial heat, warming radioactive isotopes and higher surface area compared to its volume aiding cooling. But this ebbing of the planet's geological fervour is not just written into the history of its individual volcanoes and volcanic provinces; as we have seen on the Earth and its moon, everything connects, and the legacy of Mars's smaller size and faster internal dulling pervades the planet's entire surface evolution, contributing to its delivery as the frozen barren desert that it is today.

When we compare the environmental evolution of different worlds like Earth and Mars, we need to amplify the conclusions that we draw from studying specific aspects of their behaviour. To make inferences about broad planetary-scale processes we have to integrate what are often local lessons over a world's full geography. What is more, these exercises require that we take deductions from measurements made at particular geological moment and then imagine them playing out over the almost incomprehensible billions of years of our solar system's deeper time. Our capacity to view a planet as an interrelated system and to use these broad brushstrokes to compare the environments on two different worlds is predicated on being able to see these connections here on Earth. This ability to view a world as a system is facilitated by our growing ability to 'see' our planet on a scale no longer limited by individual experience, with news travelling globally almost instantaneously and satellites scanning the Earth from orbit. But it turns out that earlier than many of these technological developments, as we shall explore next, the altitude of the Andes played at least a small role in this shift in human thinking. For humans to conceive of a planet's environment on the scale of its whole globe as opposed to more localised experience takes a shift in perspective and thinking as well as the development of the technology enabling us to see further.

We have seen over previous chapters how our appreciation of volcanism has shifted over the millennia from considering it a local phenomenon to a global system of recycling tectonic plates, huge volcanic aerosol clouds that can stretch to cool the globe and wafting carbon emissions balanced by weathering. Technological advances like undersea telegraph cables and, more recently, being able to see our world from the outside in the era of satellites and space missions have been transformative in this zooming out of our world view. Sending probes – and maybe one day humans – to other planets might be seen as the culmination of this reimagining of environment in our ever-strengthening ability to make comparisons between different worlds rather than just between Earth's contrasting regions. These technological developments are relatively recent, but as is often the case, shifts in how we think and see the world around us are driven by great thinkers as much as advances in electronics, aerospace or instrumentation. Long before high places like the Atacama were used as vantage points for great observatories to look out at the stars, human travel to elevated altitude seems to have played a role in shifting our gaze to see a bigger picture.

This was certainly the case for the hugely influential German polymath Alexander von Humboldt, who fundamentally changed our conceptualisation of nature from the local to the global scale. This change in viewpoint for him seems to have been tied to his own journey into the Andes and his own struggle with ascent and the thin air and harsh conditions of these high mountains.

Born in Berlin and educated in Göttingen and at the Freiberg School of Mines, Humboldt had long dreamed of travelling beyond Europe in the tradition of scientific explorers like Sir Joseph Banks who he had met in London. After many setbacks he finally set sail from Spain in 1799 at the age of almost 30, bound for the Canary Islands, Caribbean and the Americas. In 1802 his travels led him to Ecuador and an attempt to climb the 6,268-metre peak of Chimborazo volcano. By this stage Humboldt had climbed or attempted to climb many volcanoes including Teide in the Canary Islands and Cotopaxi while based in Quito. But Chimborazo was then thought to be the highest summit in the Andes and there was strong curiosity about Earth's highest peaks at this time. As he explained later, in 1853, '[w]hat appears unreachable exerts a mysterious pull.'

The challenges of the ascent lived up to this 'unreachable' billing. The local guides abandoned Humboldt and his companions at around 4,700 metres' elevation. After much further exertion, he and his group ended up on a precipitous ridge known in Spanish as the *cuchilla* (knife-edge). The steep and crumbly terrain forced them to proceed on all fours and the razor-edged rock fragments sliced into them, lacerating their hands. All of them succumbed to the altitude with serious nausea, dizziness, difficulty breathing and bleeding from their gums and lips. They were mostly in fog, which Humboldt described as like being 'trapped inside an air balloon'. When finally the fog lifted, it revealed both the 'grave, magnificent sight' of Chimborazo's dome-shaped summit now very close, but also their way blocked

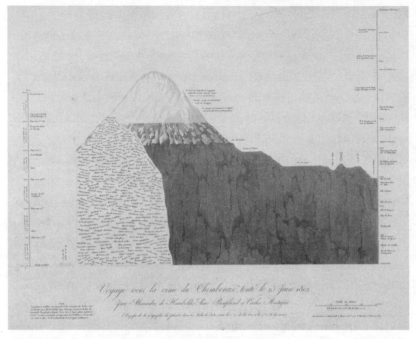

Alexander von Humboldt's depiction of Chimborazo volcano, Ecuador, with relative height scales on both sides of the image and the left part of the mountain annotated with the plant species living at different elevations. From Humboldt's Atlas, *1815.*

by a deep ravine. The barometer's mercury showed that they had reached an altitude of about 5,900 metres (although this exact height has since been questioned), but reaching the mountain's top would remain elusive.

Despite the disappointment of the unattainable summit, Humboldt's view of the world from Chimborazo was formative for him. Towards the end of his life, he often reflected on understanding nature from 'a higher point of view', from which connections could be drawn. This take on the world was strongly influenced by his experiences high on Chimborazo's slopes, maybe more than a little addled by altitude. From here, with 'a single glance', he saw the whole of nature laid out before him. Humboldt's contributions to human thought are diverse and manifold but often at their heart about making connections. He invented isotherms, the contours joining places of equal temperature that we still use on weather maps today. He conceived of the vegetation and climate zones that we use to systematise the variations in the biology and weather that stretches across the globe. He saw associations everywhere and began to see nature woven together, interconnected on a planetary scale.

In the first volume of his opus *Cosmos*, published in 1845, he took readers on a journey from outer space to Earth's inner core. He examined subjects like climate science and stretched beyond a focus simply on data, such as temperature and weather, to consider the 'perpetual interrelationship' between atmosphere, oceans, landmasses and, of course, Earth's biology. Humboldt's way of seeing and experiencing the world through which he travelled and his ideas of interconnection influenced many important thinkers in his time, among them Darwin, Goethe, Germaine de Staël, Thomas Jefferson and Simón Bolívar. Like many pioneering scholars the legacy of Humboldt's ideas echo onwards through human thought. Fields such as geosciences, geography and ecology draw daily on this vision of global interrelation, but we can now measure, analyse and simulate the world in ways far beyond what Humboldt could aspire to.

Although not formulated by Humboldt, many of his principles of connectivity are embodied by the idea – explored in the last chapter – of Earth's long-term atmospheric carbon budget and its broad long-term balance between the inputs of volcanic and tectonic degassing and removal by rainfall and reactions with the crust's rocks over geological timescales. On Earth we use this model to explain how temperatures have stayed largely within the bounds of liquid water, giving time for life to evolve in all its complexity. The alternative pathways followed by our planetary neighbours allow us to test our understanding of these balances with different planetary parameters and consequent outcomes. As discussed earlier, evidence from study of Mars's surface shows us that it was once wet, but is now a frozen desert. Further from the Sun than Earth, Mars would need a stronger greenhouse effect to keep it even as warm as our world. But the waning rates of volcanism due to its smaller size have meant that the replenishment of its atmosphere by volcanic degassing has not kept pace with the loss of carbon gases sequestered by reactions into its crust or eroded into space. In fact, as well as its diminished planetary degassing, Mars will also lose what atmosphere it accrues more readily to space than Earth due to its lower gravity and lack of a significant global magnetic field, at least in recent times. Like the Moon, its smaller size and shorter-lived internal geological activity has promoted a path to a cold and barren environment, inhospitable for life.

To sustain Earth-like conditions, with plentiful liquid water and the chance of evolving life as we know it, a planet appears to require the maintenance of a substantial atmosphere for billions of years. This needs persistent planetary outgassing significant enough to replenish and maintain a world's gassy envelope. Mars and the Moon were too small to stay geologically alive and keep these exhalations going, but our other celestial neighbour, Venus, is another matter. Venus has a similar size and density to the Earth and so should have a far greater chance of still being geologically active and with a significant atmosphere. But where Mars is cold and its atmosphere thin and frail, Venus has evolved to the other extreme. Here the atmosphere is thick, with pressures about 90 times greater than Earth, an equivalent weight to that experienced about 1,000 metres below our oceans' waves. And as well as being dense, Venus's atmosphere is a toxic greenhouse cocktail of carbon dioxide and sulphuric acid clouds that maintains its searing surface temperatures at around 500°C. These conditions are hot enough to melt lead. There is, perhaps fortunately, nowhere on Earth that we can go and imagine ourselves as being in an environment that is anything like the surface of Venus. If there were it would certainly not have spawned the backpacker bars, hostels, restaurants and hotels that make up San Pedro de Atacama, the main town in one of science's go-to Earth-based Martian analogues.

All these factors make Venus a hard planet to study. Although one of the easiest objects to spot in the evening sky, Venus's thick clouds shroud its surface in mystery. Everything we know is from space missions, with important early launches of fly-bys, orbiters and landers in the 1960s, '70s and '80s by the Russians and Americans. Venus's permanent unbroken cloud banks reflect and scatter the types of light that we can see and little of the light from the Sun gets through to the surface. Life on Venus's rocky terrain would, if we stretch ourselves to imagine that there is any possibility that it might exist, live in a perpetual state of overcast gloom, illuminated by occasional bouts of lightning. Even less sunlight finds its way back out through the clouds, frustrating chances of getting information about its surface by detecting reflected rays from the Sun via orbiters as we can for Earth, Mars or many other planetary bodies. We do have some photos from Venus's surface, from the Russian Venera missions, but in the extreme heat and crushing pressure, lander missions are difficult and short-lived. *Venera 13* holds the survival record on the Venusian surface lasting a still rather brief 127 minutes before succumbing to the harsh conditions.

But using sunlight is not the only way that we can reach out beyond ourselves for information. In 1989, NASA launched the *Magellan* spacecraft armed with a radar system to penetrate Venus's clouds and map its surface. For over four years, *Magellan* orbited the planet pinging radio waves down to the surface in strips or swaths and gradually building up a map of the shape, temperature and brightness (radar brightness, that is) of the surface. We continue to find more in the *Magellan* data even several decades later, but right from the early analyses, the mission revealed a planet covered in volcanic material. Some features are variants on familiar earthly landforms: vast lava plains and clusters of small lava domes and large and small shield-shaped volcanoes abound, along with isolated calderas: proud features on the bald surface.

Other features were unfamiliar, with the radar greyscale giving them a further strange, almost ghostly quality. Unusual formations apparently unique to Venus emerged from the data: coronae (concentric circles of fractured rock); novae (bright fractures streaking out from a centre), and arachnoids – a combination of the two, like giant spider's webs. These images and the surface data returned from the few successful lander missions leave little doubt that Venus is a volcanic world. The dearth of impact craters also hints at a relatively young planetary surface – 700–800 million years old – and possibly a global-scale volcanic resurfacing event where the whole surface was renewed in a geologically short period, although the details are still unclear.

Despite this pervasive evidence of past volcanism, whether and to what extent Venus is volcanically active today remains an open question. Recent measurements suggest lava flows of different ages and transient thermal emissions that appear and disappear on a timescale of days to months. The persistence of the sulphuric acid clouds in the atmosphere and recorded variations in the cloud-top concentrations of sulphur dioxide gas might speak of atmospheric sulphur injections during volcanic eruptions. Perhaps most persuasively, re-analysis of the *Magellan* data published in 2023 spotted changes in the shape of a flank vent on Maat Mons, one of Venus's largest volcanoes, between February and October 1991 which could be caused by active lava on the surface or the subterranean shifting of magma. Forthcoming missions to Venus will hopefully shed light on this issue, but given its comparable size and therefore thermal history to Earth, it seems more likely than not that its volcanoes are still in some way restless and venting their noxious fumes into its heavy atmosphere.

In contrast to Mars, it therefore seems that for Venus a lack of volcanic inputs to the atmosphere has not been an issue. But, as we have seen, this has not led Venus's environment on a pathway to becoming Earth-like. Venus's blistering heat and crushing atmosphere again illuminate how important planetary-scale balance is to make a palatable world. On Venus, volcanic emissions of carbon dioxide and other chemicals from the planet's interior have been able to build up unchecked by drawdown via the weathering sink we see on Earth. This appears to have largely resulted from its position closer to the Sun than Earth bathing Venus in stronger radiation than our home world. Over time, this has broken apart water molecules in the Venusian atmosphere, allowing the light hydrogen atoms to escape its gravity and fly off into space. Gradually, this has effectively stripped the planet of its vital water. Without water, rain cannot form. Without rainfall to remove it, carbon dioxide has accumulated in the atmosphere, causing a runaway green-house effect and leading to the extreme surface conditions that so challenge its exploration. While the early waning of volcan-ism on Mars and the Moon has contributed to their freezing, thin-aired inclemency, the unmoderated volcanic breath of Venus has helped to push it to the other extreme.

Despite Earth's current crisis of rising temperatures, there is little suggestion that we are on a path to a runaway greenhouse effect like Venus, at least in the foreseeable future. Things will be different in the longer-term as the Sun continues to evolve through its life cycle. In about a billion years' time, as our star brightens, a runaway greenhouse effect might become inevitable as Earth dehydrates like Venus. For now, though, the examples of the Moon, Mars and Venus should highlight for us how spe-cial our world is. Earth is near enough to the Sun to feel its heat but not so close as to have its water stripped away. Its magnetic field shields it, at least to some extent, from the erosive solar wind and it is of sufficient size and with the right chemistry to maintain volcanism, keeping its atmosphere replenished. The combination of tectonics and water harmonise to produce a cycle of carbon dioxide drawdown and emission that keeps it all in some sort of balance. Where Humboldt once had a vision of Earth's nature laid before him with 'a single glance' from altitude, we can now turn to the heavens and by looking out can see even more about the connections that sustain our world and have done so through all the eons it took for our planet to arrive at this point in time.

But it is not only by looking to our nearest neighbours that we learn new things about Volcanoland. As we gaze ever further into space, our observations stretch our definitions and understanding of volcanism ever further. Although completely conclusive evidence for current volcanic activity on our nearest neighbours still eludes us, we have, in fact, seen active volcanism on another world and it tested our wider theories of the origins of the heat held within planetary bodies. In 1979, the *Voyager 1* space probe made a fly-by of Jupiter and sent back images not only of the giant planet but also its moon Io. Linda Morabito, a young engineer working as part of the *Voyager* team at this time, was charged with image processing for navigation purposes. The mission scientists on the floor above her at NASA's Jet Propulsion Laboratory in California had already revealed some of the breathtaking colour images of Jupiter and Io that the probe had returned. In an interview with BBC *Sky at Night* magazine published in 2020 Morabito remembered thinking that Io's sulphurous hues of black, orange, yellow and blue made it one of the strangest objects in the solar system, looking something like 'a mouldy pizza'. The navigation images were in black and white and needed routine processing to calculate the probe's exact position. One morning, working on low-priority incoming data, she tweaked the brightness contrast on one of the images and, to her surprise, a gigantic umbrella-shaped plume snapped into focus. By the end of the day, alternative explanations were excluded and the team was confident that this was a volcanic fountain. Morabito remembers that brief time very early in the morning that she had alone with the image as particularly special. In her words, those moments 'were the only ones that belonged to me in terms of this discovery, because the science is more important than any one individual. It was the stuff of dreams, because I knew I was seeing something that most likely no human being had ever seen before.' This pale grey smudge against the blackness of space was our first direct evidence of another planetary body being geologically 'alive' in the here and now.

Io is similar in size to our own Moon and so the predictions had been that it would have lost enough of its internal heat to yield a geologically dead world something similar to our familiar night-time companion. The discovery of a world of such active volcanism hidden on this diminutive entity was surprising given the then broadly accepted framework governing the thermal evolution of rocky celestial bodies we have already discussed for the Moon, Mars and Venus. But, just months before the *Voyager* images beamed back to Earth, a study led by the University of California, Santa Barbara had predicted another source of energy likely to play a significant role on Io. Io is a similar distance from Jupiter as our Moon is from Earth, but Jupiter is over 300 times more massive than our home. On Earth we feel the gravitational pull of our small satellite each day in the ebb and flow of the tides. On Io, the colossal mass of Jupiter stretches and squeezes the very rock that it is made of. The rhythmic tugs of its two neighbouring moons Europa and Ganymede (a phenomenon known as orbital resonance in celestial mechanics), add to this constant gravitational kneading, and the energy from this constant deformation keeps Io's insides perpetually hot. The heat generated from the constant contortions of Io's rocky sphere fuel the persistent volcanism caught on camera by *Voyager 1* and make it, by some measures, the most volcanically active body in our solar system. On some worlds, being this geologically alive would lead to at least some sort of atmosphere, too, but on Io, Jupiter's strong pull strips away the emitted gas and keeps its sulphur dioxide-dominated atmosphere thin and fragile. Although Io's very active volcanic exhalations have not made it in any way Earth-like, Morabito's discovery and its explanation via tidally induced magmatism

An image of a high plume from Io's Tvashtar volcano taken in 2007 by NASA's New Horizons *spacecraft. In 1979* Voyager *captured an image of a similar plume produced by Io's volcano Pele.*

demonstrated that there were more forms of volcanism in the heavens than those dreamed of in our Earth-bound philosophy.

Further out in the solar system, a different type of volcanism again is in play. In 1989, *Voyager 2* flew past Neptune's moon Triton and captured images of plumes of nitrogen gas and dust blasted about eight kilometres above the surface into its thin, hazy nitrogen-rich atmosphere. Triton is about 30 times more distant from the Sun than Earth. Its surface has temperatures of -235°C and is covered in a crust of frozen nitrogen overlying inner layers of ice, and perhaps an internal briny ocean, inferred to cocoon a core of rock and metal. Examination of the images beamed back has shown a world of smooth volcanic plains – the lack of impact craters telling of a geologically young surface – marked, too, by the pits, mounds and ruffles of familiar Earth-like processes like faulting, a range of volcanic eruption styles and doming of the crust by buoyant material beneath. But on Triton these processes play out in alien ice rather than rock. Triton's icy volcanism (sometimes known as cryovolcanism) is thought to be driven by 'magmas' of nitrogen, water and ammonia rather than the molten rock dominant for Earth's volcanoes. The processes driving volcanism on Triton remain enigmatic. The moon's internal heat might play a role but there is another hypothesis that the energy might be captured directly from the Sun and harnessed through Triton's transparent crust of nitrogen-ice via a solid-state version of the greenhouse effect that we more regularly encounter to describe the energy balance of atmospheres.

Evidence for other forms of cryovolcanism across the solar system have gradually accumulated with further space exploration: from water-vapour plumes (cryogeysers) pushed from the smooth icy surfaces of Jupiter's moon Europa and Saturn's Enceladus, to possible cryovolcanic features on Saturn's moon Titan, the dwarf planet Pluto, and Ceres, the largest object in the asteroid belt. It seems as if the more we probe the solar system, the more our theories of volcanism must be adapted and expanded.

Since the first detection of exoplanets orbiting stars other than our own Sun in the 1990s, our knowledge of other worlds has moved into a greater diversity of scenarios than could be imagined purely based on our solar system. In 1995, Michel Mayor and Didier Queloz of the University of Geneva announced the discovery of an exoplanet orbiting a solar-type star. The importance of this breakthrough for science, and also human imaginings about the universe, was later recognised with the 2019 Nobel Prize in Physics. Since then, more than 5,000 confirmed exoplanets have been listed, with that number increasing all the time. The field has moved from capabilities mainly focused around detection of massive planets orbiting very close to their parent stars, to the discovery of numerous small exoplanets some perhaps even with the potential to harbour life. New missions like NASA's James Webb Space Telescope, launched on Christmas Day 2021, will extend our capabilities ever further.

Encoded in the photons that we capture from these far-away realms is information about the chemical make-up of their atmosphere, and if the atmosphere is sufficiently transparent, their surface properties. We have already discovered many astonishing new worlds. The Kepler-16 binary star system hosts a Saturn-sized planet that orbits both stars. 55 Cancri e (eight times the mass of Earth) takes less than 18 hours to complete an orbit of its host star and its surface is possibly entirely molten magma. The close-by (a mere 40 light years or 380 trillion kilometres away) TRAPPIST-1 planetary system has seven Earth-sized planets around a cool dwarf star. Three or four of these might be promising for potentially habitable conditions. As our detection capabilities increase, studying these new worlds might lead to new understanding of the Earth. They will allow us to test more scenarios of planetary evolution and how it varies dependent on factors like a planet's size, composition, distance from its star and presence or absence of a moon or moons. So far, the best-characterised exoplanets have ages from 1.4 to 11 billion years, offering the tantalising possibility that they might allow us to peek backwards or forwards through our own planet's timeline too.

As the data compiles this new cabinet of wonder composed of alien worlds scientists must stretch their minds to interpret what it all might mean, especially in terms of habitability. To understand this new cornucopia of planetary diversity we will need to draw ever more on what we understand about the planet that we know best, namely Earth. This includes evaluating what its spectral signature would look like viewed from another solar system. This is one way to test whether or how we might recognise worlds like ours across the vast expanse of the heavens. As we have explored in this and previous chapters, volcanism and outgassing play key roles in influencing a planet's atmosphere. As part of planetary-scale processes like Earth's long-term carbon cycle, volcanoes can play a crucial role in stabilising a planet's climate that might make evolution more likely to occur. But understanding volcanic emissions also plays a small part in helping us in our quest to identify potential locations that might host extra-terrestrial life among the plethora of newly discovered exoplanetary worlds. Some molecules, like methane, which might be interpreted as biosignatures if detected in an exoplanet atmosphere, can also be abiotically generated – with origins not associated with living cells – by processes including some involving volcanism and so this source must be ruled out before we re-evaluate our solitude in the universe. Characterising the atmospheric inputs from Earth's volcanoes, even the remote and inaccessible ones, yields significant insights into what we might expect to see as we look at ever-wider varieties of other worlds. Trips like mine to the Atacama to climb the prodigiously venting Lascar in 2003 are part of this ongoing global effort.

Although we drove to above 4,800 metres elevation, climbing the final 700 or so metres to reach Lascar's lofty crater rim carrying the equipment to measure the gas and particle components of the plume remains among the hardest physical feats of my life. Hiking at altitude is totally different to sea-level exertion. It is a slow, deliberate plod. Once you are in the rhythm it can feel something like a meditation in that you can think of little more than your breath and putting one foot in front of the other: breathe, step, breathe, step, breathe. Fortunately for us, although it was certainly a struggle, we did not bleed from the gums and lips like Humboldt and his crew. Again, in contrast to Humboldt's trial on Chimborazo, there was no cloud-deck for us to emerge from. The day that we climbed was staggeringly clear. But the dizziness that he reports is certainly a sensation that we shared. I have to admit that my memory of the summit is a little hazy, likely from the altitude, but I do remember looking at my electronic altimeter, rather different to Humboldt's mercury contraption, and noting that we were at half the atmospheric pressure of sea level. At this altitude, half the mass of the atmosphere was below me. There is a photo of me slumped on my back in a cleft of an enormous breadcrust bomb (a dense chunk of rock thrown from the volcano with a characteristic cracked outer surface that solidified in flight and fractured by contraction, expansion or impact), looking as if I had broken it myself, and I can well recall how heavy my body felt from the exhaustion of the climb and the sparse oxygen. I remember, too, gazing out into the almost impossibly immense blue of the heavens with a certain peace, a detached observer of the world below, which maybe draws some parallel with Humboldt's sense of epiphany in the altitude. Sadly, though, this altered mental state did not gift to me even a minuscule fraction of the

The author resting on a breadcrust bomb near the summit of Lascar volcano in 2003

insights that were to flow from him.

When I stood gazing down on the world from the top of Lascar, listening to the hum of the pumps drawing the volcanic gas and particles through our samplers, it was still less than a decade since the detection of the first exoplanet and science had only very recently made the first measurements of exoplanet atmospheres. In 2001, lines in the light detected by the Hubble Space Telescope from a planet nicknamed Osiris, 220 times the mass of Earth and 159 light years away, proved to be the finger-print of a sodium atmosphere swathing the planet. Despite the challenges of the measurements in 2003, hauling the batteries and filters up Lascar allowed us to determine the ratios to sul-phur dioxide of the acidic halogen halides like hydrochloric acid and particles like sulphuric acid haze fuming from the summit. Determining sulphur dioxide fluxes, using a spectrometer like at Masaya, involved stomach-churning rides along the bumpy dirt roads while trying to read the laptop screen that charted the algorithm's counting of the molecules in a slice or cross-section of the drifting gas trail from the crater. Often the roads ran out before we could get under the plume, and the land mines in the area, left over from a previous border dispute, meant that we dared not take the four-wheel drive too far off the beaten track – in other words, where we could see previous tyre marks or footprints. But we managed and put together the most complete chemical inventory of the volcanic emissions at that time showing Lascar to be a substantial source of acidic gases to the rarefied desert atmosphere.

Five years later, we compiled these rare measurements at Lascar with other hard-won data from the international volcanological community regarding the emission of hydrochloric acid to make a best estimate of the global flux of this noxious gas to our atmosphere from volcanism. These estimates have since fed into efforts to model whether the James Webb Space Telescope might be able to detect the existence of biology or oceans on the planets in the TRAPPIST-1 planetary system. Other measurements from that day informed our understanding of the potential impacts of volcanoes on Earth's nitrogen cycle, again of interest when looking for evidence of signals of biology like ammonia in exoplanet atmospheres. Exoplanet research was not what motivated us to make our measurements at Lascar's summit, but it is part of the excitement of doing science that you never quite know where your work might find a use. Humboldt was a consummate explorer whose thinking was before its time in terms of how he saw large-scale connections. It is impossible to know, but I like to think that this unexpected connection between the fuming peaks of his beloved Andes ('[i] n the mountains is freedom!') and humanity's new frontier of exploration of worlds beyond our solar system is something of which he would have approved.

*

Humboldt is one of only two people immortalised on the Moon by having maria named after them. Aptly, given its namesake's passion for exploration and connection, the Mare Humboldtianum is situated between the visible near side and dark far side of the Moon, linking the seen with the shadows. Humboldt was not a fan of the idea of human expansion onto other worlds. He had seen again and again on his travels how humankind unsettled the balance of nature. He talked of 'mankind's mischief . . . which disturbs nature's order'. He was bleak about the prospects for humankind and feared that eventual expansion into space would just spread our lethal mix of vice, greed, violence and ignorance across other planets. As early as 1801, he wrote that the human species could turn even those distant stars 'barren' and leave them 'ravaged', just as we were already doing with Earth. It is hard to imagine what he would make of the data coming in regarding the status of our planet almost 200 years since he wrote those words, or the exact role that he would play in the science and debate surrounding climate change and how to mitigate its worst effects.

Earth is certainly changing very rapidly at present. We have seen in the previous chapter how human driven carbon emissions very significantly outstrip the emissions from the world's volcanoes. As long as the checks and balances of carbon drawdown by rain and reaction are in operation, in the very long term the Earth will re-stabilise and will not run away into the crushing heated hell of Venus. But what effects will these great increases in atmospheric carbon have, and what will be left once events have run their course? Here again an adventure into Volcanoland has answers as well as warnings for us. To unravel these messages we must take all that we have learnt about volcanic gases and activity and travel back into the geological record. We must remember that for our species to assume its role of dominance on this world today, vast numbers of other species had to die out. This process of life's diversification and contraction has not always been constant. Deep in geological time, as well as periods of burgeoning life, there lurk periods of great dying. Evidence links these mass extinctions with volcanism, but a type of volcanism that our species has yet to encounter. It is volcanism on a scale we should pay attention to, not only because it will one day happen again on Earth, but also because it might hold ominous analogies of the possible futures our current dependence on burning fossil carbon could unleash.

Chapter 8

Mass Extinction

*Did large-scale volcanism trigger past
extreme environmental change?*

Despite its importance, the current alarm about the state
of our planet's environment voiced across the globe is
just the latest in an increasingly urgent cacophony of concern
that has been building for centuries. We saw in the previous
chapter how Alexander von Humboldt highlighted humankind's disturbance of nature's order over and over again in the
early 1800s. Thinkers like the American George Perkins Marsh
were inspired by Humboldt's writings, adding their own data-
gathering and observations to flesh out how humans ravaged
the environment on many scales. Marsh travelled extensively,
including through Turkey, Egypt and Italy, and in his 1864
book *Man and Nature* urged his fellow Americans to learn from
the lessons of the Old World. He believed that the Roman
Empire had fallen due to the destruction of its forests, leading
in turn to degradation of the very soil that fed its people. For

Marsh, the key to avoiding a dark future was to learn from humankind's past mistakes.

Man and Nature certainly had influence. It inspired laws in the US, including the 1891 Forest Reserves Act, and informed debate about the importance of conservation into the twentieth century. However, if we look at present-day statistics, deforestation and forest degradation, as well as other wider environmental damage, in the Americas and globally, continue to take place at an alarming rate. Since Marsh's time there has been the increasing realisation of how global our species' impact on the environment can be. Emissions of long-lived gases like carbon dioxide from the burning of fossil fuels in one nation or group of nations cascade to climatic changes and acidifying oceans across the world.

Concerns about human carbon emissions and our disruption of the planet's geological balance sheet between carbon outgassing and intake, as we have seen, are also not new, and emissions from Earth's volcanoes give us a scale bar for the perturbation to the atmosphere over which we preside. In a paper published in 1957, Roger Revelle and Hans Suess of the Scripps Institution of Oceanography in California noted that projections of human emissions of carbon dioxide by fossil fuel burning in the first decade of the twenty-first century were a hundred times greater than the likely usual rate emitted by volcanoes. By returning to the atmosphere and oceans the organic carbon stored over hundreds of millions of years in sedimentary rocks, like coal and oil, within a few centuries, Revelle and Seuss wrote that 'human beings are now carrying out a large scale geophysical experiment of a kind that could not have happened in the past nor be reproduced in the future.' They continue, perhaps somewhat dispassionately, that 'if adequately documented' this experiment 'may yield a far-reaching insight into the processes determining weather and climate.'

In Humboldt's time, atmospheric carbon dioxide levels were

barely above the 280 parts per million (ppm) that characterised the pre-industrial atmosphere. When Revelle and Seuss wrote those words in the 1950s, levels were just shy of 320 ppm. By 2020, average levels were 412.5 ppm. As mentioned before, in 2019 human fossil fuel burning released over 35 billion tonnes of carbon dioxide to our atmosphere; Earth's volcanoes on the other hand emit at most about half a billion tonnes per year. Although our current 'experiment' is in many ways unprecedented, there is much to learn from the geological past. As alluded to before, our planet's volcanism today doesn't reflect the full portfolio of its ability to mobilise carbon from Earth's solid innards to its gassy envelope of atmosphere. The human experiment with carbon is unique but not without its analogues in Volcanoland.

Hidden in the ice cores, as well as the sulphate spikes discussed in Chapter 5, are tiny bubbles of trapped air. Measuring the carbon dioxide concentration of these bubbles gives us an 800,000-year record of atmospheric levels. The highest concentrations hidden in the ice are about 300 ppm. We are in uncharted territory in terms of the memories held in the compressed depths of our planet's ice caps. But it is not just ice that can hide clues about our planet's past. Formed as they are in some degree of dialogue with Earth's past atmospheres, oceans and biology, sedimentary rocks capture messages about the past state of our planet's environment and this record goes far deeper into time than ephemeral ice. In the sedimentary record we see evidence for atmospheric carbon equal to and higher than present-day levels, although we do need to look back more than three million years.

Before humans, geological forces dominated the control of Earth's carbon. In Chapter 6 we witnessed the role of volcanoes as part of our planet's broad long-term carbon cycle, a unique (according to our observations to date) and lucky equilibrium that has arisen on Earth from a very specific set of circumstances, as we have seen from the other worlds. But volcanism itself can also challenge this balance temporarily on Earth and has seemingly done so more than once in our past. We saw in Chapter 6 how short-lived eruptions, even those large enough to be classed as supereruptions, do not emit enough carbon dioxide to significantly change atmospheric concentrations. However, these powerful but nonetheless geologically fleeting events are not the only form of extreme magmatism that humans have yet to witness. As explorers have described and mapped our planet's surface, great outpourings of volcanism, floods of basalt, have revealed themselves, covering huge footprints of the landscape and telling of another type of volcanic episode unlike anything that we have recorded in our short human history. For lessons about the likely outcomes of what Revelle and Seuss term our 'large scale geophysical experiment' with the planet's tolerance to carbon, looking at these floods of lava and their consequences is one promising place that we can turn to.

Through my explorations described here thus far, I have related many instances from throughout my career that have been all about awe at the power of volcanoes: their heat, their destructive power, their ability to stain the evening sky and chill the air across the planet, their ability to maintain a stable climate, their reach across the heavens. Their dominion over our world and other worlds, tracing the eons of existence, often makes me feel insignificant. Looking into the geological past of their greatest episodes might seem a strong candidate to further invoke this sense of vertiginous reverence. However, while this emotion is certainly present, in the juxtaposition of my professional studies of the geological record and the increasingly frequent news stories about retreating ice, wildfires and extreme climate events, I find a different message too. More and more, delving into Volcanoland is a reminder for me about our own more recent power as a planetary-scale process. The lessons that we can learn from these unavoidable past natural experiments, presided over by magma millions of years ago, hold warnings regarding the gravity of the choices that we are making in the present day, now that we are the guardians of Earth's future. As we have seen from our adventures in previous chapters, too, trying to understand planetary-scale processes is challenging, but usually starts with looking at the world around us in some way. In this case, let's start by exploring the landscapes left by these huge outpourings of lava that mark these periods of frantic volcanism hidden in Earth's past.

There is a place in the northern part of the Isle of Skye in northwest Scotland where a great cliff face looms large in craggy tiers up to a flat plateau and the summit of Meall na Suiramach above. It is a long way from any present-day volcanic action, but its face exposes layer upon layer of ancient weather-beaten basaltic lava flows. I have only ever been there in the mist and rain, characteristic conditions for this part of the world, and I have not yet picked my way along the path among the needle-like pinnacles and sheer scarps up to the tabled summit above, although it is firmly on my wish list. This area has long been a playground for geologists. Scottish geologist Archibald Geikie took a keen interest in the volcanic rocks of western Scotland and how they related to each other. This northern tongue of Skye was special for him too. In 1897, he captured this landscape where the 'green terraced slopes, with their parallel bands of brown rock formed by the outcrop of the nearly flat basalt-beds, rise from the bottoms of the valleys into flat-topped ridges and truncated cones'. Lava flows have written these hills into 'a curiously tabular form that bears witness to the horizontal sheets of rock of which they are composed.' This basalt-plateau juts out in places across a band from this northern part of Skye down through the Isle of Mull and into Northern Ireland, including the famous basaltic columns of Giant's Causeway in County Antrim and Fingal's Cave on the small island of Staffa near Mull.

There is further more muted evidence for magmatism of a

Layers of ancient North Atlantic Igneous Province flood basalt flows on Mull, Scotland, taken in 2016

similar age as far afield as Lundy in the Bristol Channel. Geikie remarked on the wide areas that these plateaux embraced and the evidence that their limits 'are now greatly less than they originally were is abundant and impressive'.

Between 55 and 60 million years ago, twists of fiery lava would have stretched over the landscape of what is now north-west Scotland. In episode upon episode, these incandescent flows cooled and dulled to build the topography layer upon layer. Despite the great length of time these rocks have sat on the surface of our planet, in places you can still make out the textures telling of flow and ropey *pāhoehoe* surfaces reminiscent of fresh lava pushed out in places like Hawaii today. But on Skye the work of weather over the many epochs since their eruption has also carved the different lava layers into the tabular steps that so often plough into the wind and rain and shepherd the mists but nonetheless captivate visitors like me today.

These layered landscapes of lava, with their terraced topography, colossal horizontal steps and vertical sloping walls, were initially known as 'traps' after the Scandinavian terms (*trappa* and similar) for stairs. They were recognised as a distinct type of volcanism as early as the 1860s, and it was apparent that these were massive events. In the second half of the nineteenth century, Aleksander Czekanowski, a Polish geologist, was exiled to Siberia by the Russian authorities for his alleged participation in the Polish–Lithuanian 1863 January Uprising. Between 1869 and 1875 he surveyed the geology of the region and found valuable coal and graphite deposits. But in his expedition diaries (published in 1896) he describes the 'best results of this geological study' of Siberia as the 'discovery of a previously unknown area of igneous rocks of so large an extent that it exceeds the size of any other of its kind'.

He mapped these Siberian 'traps' along the Nizhnyaya Tunguska River and north to the Olenek River, tracing them 'throughout six degrees of latitude and 15 degrees of longitude', an area of about 1.1 million square kilometres. The extent of these vast trap lands is even more impressive from the skies. Commercial air travel has opened up new views of the world. On a flight from London to Tokyo in 2016, over 140 years since Czekanowski's expeditions of discovery, I sat mesmerised gazing at the seemingly endless Siberian Traps, the vast plateau carved by the dendritic drainage of rivers, spreading out beneath me as we passed over high above – a monumental volcanic handprint outlined on the Earth below.

These traps in western Scotland and Northern Ireland and Siberia are by no means unique globally. Near the US city of Portland, Oregon, the Columbia River drains through similar high, terraced cliffs. Here this table-land topography drapes over about 240,000 square kilometres of the landscape. Up the Eagle Creek Trail, I once hiked up into the flows along an overhung path carved into the rock of the steep valley side, waterfalls gouging out the ancient lava. The Deccan Traps blanket great swathes of India; the Ethiopian Highlands – where the Blue Nile rises – are in part traps; on the border between Brazil and Argentina water plunges over the Iguazú Falls down steps cut into flow upon flow of ancient basalt, remnants of the Paraná-Etendeka large igneous province. There are many more examples beyond these few with all of Earth's continents showing at least some evidence of this mighty type of volcanism tiering forth deluging ancient sectors of their landscapes deep in the past. But they are not only found on land. Some form vast plateaux that carpet the ocean floor. One of the largest of these preserved, the Ontong Java plateau, covered about one per cent of the Earth's surface when it erupted. It now makes up a huge area of shallower ocean in the south-western Pacific, north of the Solomon Islands.

Geikie built trap topography into his system of volcano classification, referring to it as 'Plateau type' volcanism. In 1932, George W. Tyrell, another Scotland-based geologist, coined the term 'flood basalts' to describe the way in which 'the topographic features of the overflowed region have been swamped by lava as by a flood'. Since the 1990s they have been known by their more general name of 'large igneous provinces'.

In some places like Siberia and the Columbia River, the full footprint of these large igneous provinces can still be appreciated, while in others they have been torn apart as oceans opened or supercontinents rifted. The table-like plateau of northern Skye was once much closer to other remnants now as far afield as the Faroe Islands, Iceland and east Greenland. These flood-basalt fragments, known collectively as the North Atlantic Igneous Province, were ripped asunder as the north Atlantic opened. In northern Africa, South and North America we find rocks from a large igneous province known as the Central Atlantic Magmatic Province – continental jigsaw pieces

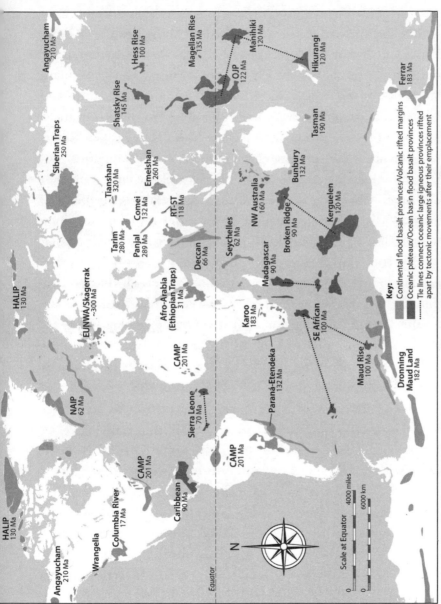

A map showing the global distribution of basaltic large igneous provinces erupted over the last 320 million years of Earth's history with ages indicated in millions of years before present (Ma). Abbreviations: **CAMP**-Central Atlantic magmatic province; **EUNWA**-European, northwest Africa; **HALIP**-High Arctic large igneous province; **NAIP**-North Atlantic igneous province; **OJP**-Ontong Java Plateau; **RT-ST**-Rajmahal Traps-Sylhet Traps.

scattered around the ocean's margins and into the continents. In some places these outcrop in relatively remote places like the Atlas Mountains in Morocco; in other places they are embraced within the infrastructures of modern urban living. The striking Palisades sill that forms prominent cliffs along the western bank of the lower reaches of the Hudson River is associated with the Central Atlantic Magmatic Province. I enjoyed a rain-flecked view of it through the train window riding from Tarrytown to Marble Hill on a recent trip to New York. Rewinding the tectonic dance of the continents, these pieces fit back together and reveal their shared origins as part of the ancient supercontinent of Pangea. The rocks from more than 24 of these great floods of magma have left their indelible impressions on Earth's surface over the last 275 million years. The remains from others – from further back into deeper time – are present, too, but the record becomes more and more patchy with ancient rocks lost to subduction, weathering or mashed up by the shifting plates.

The known large igneous provinces vary in size but most spewed out millions to tens of millions of cubic kilometres of lava, with more magma left behind as saucer-like lenses or sills and pillared intrusions or dykes within the crust. To give a sense of the scale of these huge volumes, a cubic kilometre would cover the whole of Greater London in a crusty layer of lava about 65 centimetres deep. These events overran the landscape with quantities of lava millions of times this volume. The origin of this massive melting of the mantle is still debated, but a plausible explanation is vast plumes of hot mantle material ballooning up within the planet. Their origins could be as deep as the core–mantle boundary, and they would have been similar, if likely larger, to what we image using seismology feeding the East African Rift today, as described in Chapter 6.

Human history holds no records of large igneous province volcanism and its aftermath. In fact, our investigation of these extraordinary regions of rock shows that their most recent episodes pre-date the existence of our species on the planet. To learn more about them we must piece together evidence from many sources but central clues come from reconstructing the timing, duration, intensity and style of their volcanic eruptions from the rocky remnants that they have left behind. As we saw on Santorini, piecing together past volcanic events from rocks that have sat on the Earth's surface for just a few thousand years is challenging enough. This is only magnified for the lavas left from large igneous provinces that have been worked over by our planet's restless tectonics and weather for tens of millions of years or far more.

The best understood is the youngest, erupted in what is now the area surrounding the Columbia River in the US north-west and known as the Columbia River Basalt Group. It defines the local landscape, and when I was there, I hiked up amid the mossy trees and cliffs and watched the water pour over its basalt steps. The scenery is breathtakingly beautiful and impressive, but it is nonetheless also one of the smallest known examples of a large igneous province, weighing in at only about 0.25 million cubic kilometres of lava. The Deccan Traps, the Central Atlantic Magmatic Province and the Siberian Traps are all ten or more times larger when it comes to sheer erupted volume.

These freshest of large igneous province rocks, chiselled out to form this section of the Columbia River gorge, hold crystals that can be dated by measuring the decay of radioactive elements within them. The basalts around the Columbia River yield ages from as old as 17 million to as young as six million years. But this wide age range hides periods of far greater volcanic ferocity. Detailed mapping and dating of the flows suggest that over 70 per cent of the province's volume was erupted during a much shorter 400,000-year period around 16 million years ago. During this peak in activity, estimates are that volcanic outpourings in the area would have been at rates comparable to or greater than those over the vastly shorter eight-month Laki eruption that devastated Iceland in the 1780s as described in Chapter 5. In fact, during its most frantic period, the rate of volcanism from the Columbia River Basalt Group potentially dwarfed all other volcanism, pushing and spewing out rock onto Earth's land at the time, added together. The more violent pulses in activity appear to have lasted decades and beat out the rhythm of the crescendo and diminuendo of the province's more extended lifetime.

For the provinces older than the Columbia River basalts, dating of individual flows is more difficult, but we can get a sense of the general timing of volcanism overall. As well as the timescales defined by the ticking clocks of radioactive decay, geologists use other clues, such as those from the weak magnetic signatures left in the lavas that tell of ancient reversals in Earth's magnetic poles. These intricate measurements fit these basaltic floods into Earth's geological timeline, indicating that the Deccan Traps erupted around 66 million years ago, the Central Atlantic Magmatic Province about 200 million years ago and the Siberian Traps 250 million years ago. Piecing together the timing of all the different flows is challenging, but our best efforts tend to support the findings from the Columbia River of a shorter (roughly one-million-year) more intense period of activity sandwiched between longer waxing and waning phases. Getting more detail about the peak eruption rates during the emplacement of these older provinces is tricky, but the huge volumes of lava mapped and tallied on Earth's surface today yield eruption rates that are highly significant on a planetary scale, even when averaged over the million-year spans of their main activity. Lessons from the Columbia River and Laki lead us to hypothesise that within these periods eruption rates would like have been far higher at times, likely dwarfing those we see today from all the world's subaerial volcanoes added together for substantial stretches. But we know that there would have been quieter interludes too. For example, in the Deccan, we find red clayey soils that had time to form between the flows, telling of periods of relative calm interspersed with episodes of great ferocity.

Although the exact details of the eruption style of the volcanic episodes that drummed the beat that built these great stepped landscapes is hard to reconstruct from their ancient rocks, some broad lessons have long been very clear. Large igneous provinces are a different mode of volcanism from the swift cataclysms of huge but short-lived eruptions like Tambora or 'super eruptions' like Yellowstone or Toba. The deposits from Tambora or Toba speak of far-flung ash, Plinian columns punching into the stratosphere and huge outflows of swiftly dealt pyroclastic flows. The vents that spewed the lavas of flood basalts are rarely preserved, but where they are we do not see the same evidence of vast explosive activity. Fire fountains would have, at least periodically, blazed out of fissures splitting the earth and feeding the snaking flows, but the eruptions of these runny, water-poor basalts present a very different image from our usual concept of volcanic catastrophe. The power of flood basalt volcanism is not as much in its immediate global impact over weeks or years but rather its persistent and prolonged action. Where events like Toba provoke acute disruptions to the Earth system, large igneous provinces are like a chronic condition playing out over a sufficient duration to be significant when measured by the yardstick of geological time.

Eruptions like the fissure eruptions characteristic of Iceland and Hawaii are our best present-day guide to the style of flood-basalt eruptions, if on a much reduced scale. We can use our experiences at these locations to imagine a blackened landscape, curtained by the raging incandescence of fire fountains. Red-hot lava flows would have laced their way through the terrain, torching any vegetation in their path. Billows of gas and aerosol would have lifted with the heat of the fountains, perhaps sometimes reaching above ten kilometres up into the atmosphere, with more haze drifting from the flowing lava downstream. The ground would have throbbed with the deep pulse of volcanic explosions. The scene would have been apocalyptic and the air would have been acrid with volcanic fumes. The accounts of the Laki eruption give us a better guide in terms of the intensity of large igneous province volcanism than anything in the present day, and Laki is usually cited as the closest historical analogue to these epic periods of magmatism. But pushing out only 14 cubic kilometres of lava flows, Laki is still an absolute minnow in comparison to even a small flood basalt.

Nonetheless, the chronicles of devastation from 1780s Iceland can give us some indication of the environmental fallout of 'Plateau-type' volcanism. As during the eruption of Laki, the outpouring lava cloaking the landscape would have been a local catastrophe, but the noxious gas and aerosol would have poisoned the air, water and soil over a much greater area than that blackened by the flows. Strange hazes and weather would have plagued the surrounding region, making it challenging for local biology to thrive, or maybe even to survive. As we saw earlier, Benjamin Franklin measured the shadow of Laki's haze in the weakened sunlight thousands of kilometres from Iceland in France. Scaling up from Laki's impacts to that of a large igneous province, about a million times the volume and duration, might seem a challenge, but is important to attempt in our quest to understand them. Estimating the total quantities and fluxes of gas emitted from ancient large igneous provinces is, like reconstructing eruption rates and styles, fraught with difficulties. As described before, sometimes we get clues about how much gas the magmas carried from deep by analysing the tiny blebs of glassy rock trapped in the crystal cargo of the flows and carried up like minute time capsules from Earth's depths. But in these very ancient rocks these messengers can be tricky to find and, as for modern volcanism, they can give particularly poor information about a magma's deep carbon contents.

Even so, with the very basics that we know about magma and its ability to transport gas from inside the planet to the surface, we know that the fluxes of important gases like carbon dioxide, sulphur dioxide, and hydrofluoric, hydrochloric and hydrobromic acids would have been huge. Other sources of gas might come into play, too, for a magmatic system of this scale. Spanning from the mantle to the surface, these huge subterranean catchments have the potential to tap ancient carbon supplies tied up in the brittle mantle under the continents, in a similar way to what is proposed for the East African Rift today. Sometimes, also, as in Siberia, the magmas cut through coal seams or other deposits on their journey to the surface. Magmatic heating of these deposits can release gases like carbon dioxide, again adding to the potential output of vapours from the planet's innards to the surface environment associated with these colossal periods of heightened volcanism.

Although it is very challenging to estimate the total amounts of different gases released from events this old, we can use what we know to at least evaluate their most likely scale. Estimates of total carbon dioxide emissions over the million-year-scale lifetimes of these basaltic floods are in the region of tens to hundreds of trillion tonnes. Human carbon dioxide emissions since the industrial revolution come in at more like a few trillion tonnes so far, but we've been at it for a far shorter period of time. Fluxes of carbon emission from a large igneous province were, for the most part, likely much lower than those since the invention of the steam engine. Estimates of varying emissions rates over the very long lifetimes of these provinces are even harder to make than the totals, but one recent study put the maximum emission rate during the Siberian Traps at around 18 billion tonnes of carbon dioxide per year: about half the current carbon emission rate from our burning of fossil fuels. Sulphur and acidic halogen gases are emitted in lesser but nonetheless significant amounts.

Given the enormous scale of these large igneous provinces and the historical testimony of Icelandic devastation and strange regional-scale weather from our shorter-lived but closest analogue, the Laki eruption, it makes sense to ask what impacts these vast events doled out for our planet's environment during their tenure. We can certainly learn a lot by analogy to more recent or present-day volcanism. In previous chapters we have seen how we can build up a picture of recent and current eruptions and the aftermath of their flows, ash, gas and particles for the environment. But, given our (fortunate) lack of direct experience of a large igneous province event, to fully understand the consequences of volcanism on this scale we must also turn to the deep geological record to understand how these events changed both the environment and biology present on our planet at the time of their eruption. Clues about Earth's past environments are all around us, hidden in the sedimentary rocks that outcrop at the surface and often build the ground beneath our feet. These rocks host a record of the Earth's life and times written in the fossils and whispers of past chemistry caught within their layers as they settled in sequence over the millennia, eras and eons. To look for signals of environmental fallout from large igneous province volcanism we must first consider the broader sweep of geological time and how we track the patterns of change in Earth's biology and chemistry that finally arrive at where we are today.

The Earth is almost unimaginably old, and it has not always been anything like it is in the here and now. Humanity has long pondered how the planet came to be how it is, and while myths, legends and beliefs still abound, studying the record left about its past in its rocks, along with other clues, has yielded great scientific advances in our understanding of our origin story. We have found out a lot, including how very short our history as a species is on this planet. Early in my PhD, I attended a lecture in a dusty Cambridge auditorium one cold autumn morning. It was about the geological timescale and, although I have to admit at the time being overwhelmed by the complex system of eons, eras, periods, epochs and ages, one analogy from that lecture lodged firmly in my memory. The lecturer stretched out his arm sideways and asked us to imagine that it represented the age of the Earth, with his shoulder the beginning, 4.5 billion years ago, and his fingertips the present moment. On this scale he challenged us to guess for how much of his arm human beings had existed. I thought maybe we might get at least a knuckle, some others thought maybe as much as a hand, others were savvier and already knew that if you take a nail file and draw it across your fingers once, that illustrates the proportion that human beings have graced this rock: just about 200,000 of 4.5 billion years. Our species' sense of self-identity in terms of our world should always draw upon our status as just the nail dust of geological time. Sometimes, if the day-to-day stress of living is getting to me, I will pause for a moment to remember this and reset my perspective. Here again is a sense of fear and awe akin to staring out into the vast endless expanse of a starry sky or into a roiling lava lake or imagining the great depths of the Earth right down to the core's centre in a creeping cascade beneath you.

The evolution of Earth's surface environment to become what it is today is deeply linked with the evolution of life on this planet. Early Earth would have been completely unrecognisable. It is hard to establish when life began with high confidence but it seems likely that for at least the first half billion years, or about one third of the way from my shoulder to my elbow, even microbial life was absent. By dating fossil-bearing rocks we can see that the advent of multi-cellular organisms and eventually plants and animals didn't come until much later. The 'Cambrian Explosion', accepted as a key revolution in terms of diversity of species and animal body plans in the oceans, was about 539 million years ago, or at about at the base of my fingers. It took longer still for complex life to move from the oceans to colonise the continents. Although microbial life on land likely dates back far further, the earliest good evidence of land plants and animals dates back to what is known as the Ordovician Period (485 to 444 million years ago or creeping up towards the middle joints of my fingers). Modern land ecosystems only established, about 385–359 million years ago (the Late Devonian Period), just about at my middle knuckles. It is sobering, or perhaps calming, to contemplate that for most of Earth's long history, the continents were barren places devoid of complex life, the playground only of wind, water, ice, rockfalls, tectonics, volcanism and the occasional meteorite impact albeit, after some hard to discern geological moment, presided over by microbes.

Understanding the history of life and the environment on our planet is intimately entwined with understanding its rocks. The debris of ancient life is both trapped and recorded in rocks and, in some cases (for example chalks, coals and stromatolites) is fundamental to making them. Downlands of open chalk hills dominate the landscape to the south of where I live in Oxford. Wandering the hillsides around the ancient human relics at Avebury, West Kennet and the Uffington White Horse, the white scars, exposed in the places where the grass and soil yield to the rock below, are the dusty remains of even more ancient life: Cretaceous microorganisms from long-lost oceans.

Fascinating as these pale microbe remains are, bigger types of fossil readily visible to the human eye are easier to appreciate as a sign of past life. Human interest in the fossil record runs deep. In places like France, fossils of trilobites, extinct for about 250 million years, are found drilled for use as pendants in prehistoric human cave sites. In classical times, the Greeks and the Romans began to collect fossils, consider their meaning and even use them as medicines. Sucking fossil spines and teeth or taking them as powders was given in treatment for many ailments, from bladder stones to poisoning. Learning to read the messages about our planet's story encoded in these fossil relics took humans longer and necessitated some big switches in the way we saw the world.

As in the history of volcanology, early European thinking about fossils and their meaning was strongly influenced by Christian narratives. Fossils were generally seen as relicts transported and then buried by the biblical flood. This also meant that biological change or dying out (extinction) of species was a problematic concept for western thinking prior to the nineteenth century. Much of Western society believed that God had created a world that was perfect and complete. Fossil finds such as impressions of giant nautilus-like shells were explained away as extant species simply hiding in unexplored regions of the Earth. However, by the late 1700s, thinkers were beginning to use evidence from the fossil record to make the argument for a more complex history of life and the environment on our planet. French naturalist Georges Cuvier argued that mammoth skulls found in the Paris basin were distinct from any known living species of elephant and that it was highly unlikely such an enormous animal would go undiscovered. Across the channel, William Smith was using the characteristic fossils to identify and link up the different strata across Great Britain as he surveyed the mines and canals, pulling together these threads to produce the first detailed geological map in 1815. In the first half of the nineteenth century, others including the British geologists Roderick Murchison and Smith's nephew, John Phillips, used these nascent techniques of biostratigraphy to start to sketch a systematic geological timescale, creating a new language to describe Earth's vast pre-human history.

Whether biological change manifested as species extinctions was gradual or occurring in cyclic cataclysms was also long debated. Cuvier's studies of the French strata led him to see marked cycles of devastation and rebirth in the planet's biotic

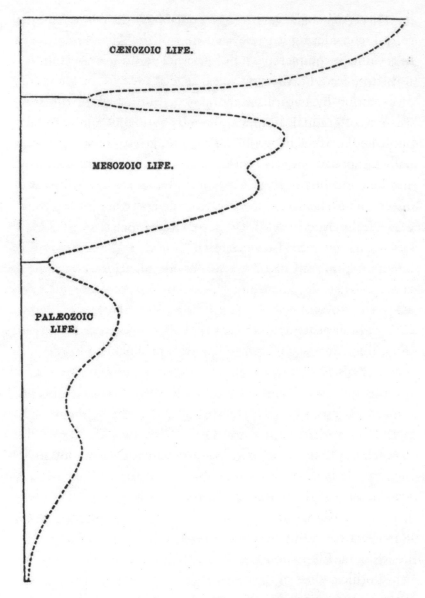

A 'continuous curve, which corresponds to the numerical prevalence of life, and represents its rise and fall' plotted by John Phillips using British fossils and published in his book Life on the Earth: Its Origin & Succession *(1860). The low point in the curve at the top of the 'Palaeozoic Life' section we would now call the end-Permian mass extinction and the dip at the top of the 'Mesozoic Life' section corresponds to the end-Cretaceous mass extinction.*

diversity. Although Darwin had mixed feelings about the fossil record, considering it to be 'a history of the world imperfectly kept' in its incompleteness, he nonetheless drew important lessons from it and extinction was a crucial process for his theory of evolution by natural selection. For him, though, the story was one of gradual disappearance. In *On the Origin of Species*, published in 1859, he wrote 'species and groups of species gradually disappear, one after another, first from one spot, then from another, and finally from the world'. In contrast, Phillips produced one of the first plots of species diversity over an extended geological period in his 1860 book *Life on the Earth: Its Origin and Succession*. Based on the fossils found in the layers of sediments making up the British Isles, he identified 'surfaces or zones of least life', great dips in 'the abundance of the forms of life in the sea' with those at both the end of the Permian Period (252 million years ago) and the end of the Cretaceous Period (66 million years ago) being the most remarkable. On the page opposite his plot of wildly unequal species diversity with time, Phillips quotes the Roman poet Lucretius: 'in a brief interval the generations replace each other and like the racers transmit from hand-to-hand the torch of life'. Phillips spent the end of his career in Oxford and in terms of his personal handing of the torch of life from generation to generation, he had what I have seen described, I hope inaccurately, as a 'very Oxford death': after a fine dinner at All Souls College, so the story goes, in illustrious company, he tripped over a mat and fell down the stone stairs. He is nonetheless immortalised in the Victorian neo-Gothic cathedral to science of the Oxford Natural History Museum, just around the corner from my office, the building of which he oversaw. His carved bust is among the science greats, including Darwin, adorning the pillars of its vaulted space.

It took another century after Phillips's unfortunate personal postprandial extinction before the matter of measuring species diversity through geological time was revisited in detail. By this time, humanity was squarely on the ramp of the information age and the rapid developments in computer power could be brought to bear on the gradual eras and epochs of the fossil record. In the 1970s, American palaeontologist Jack Sepkoski built a simple computer database of the appearances and disappearances of fossil marine invertebrates and vertebrates from the record since the time of the 'Cambrian Explosion'. In 1982, along with his University of Chicago colleague David Raup, he used this database to identify five devastating extinction events in the history of Earth's life, where the timelines of great tracts of species get cut out of existence. These so-called 'Big Five' mass extinctions include the events spotted by Phillips at the end of the Cretaceous Period that wiped out the non-avian dinosaurs, and, most severe of all, the great dying at the end of the Permian Period, when estimates show that over 80 per cent of species perished. Sandwiched between these two, 201 million years ago, is the end-Triassic mass extinction, and then further back in geological time, Sepkoski and Raup identified the late Devonian (372 million years ago) and Ordovician-Silurian (444 million years ago) extinction events.

The rocks that tell of these dramatic periods of demise are scattered across the globe and are sometimes hidden in somewhat unassuming places. From my home in Oxford, both the end-Triassic and end-Permian Periods are within easy reach. About three hours' drive takes you to the grey shingle beach at Budleigh Salterton in Devon on the UK's south coast. Here the cliffs are a deep brick red and the sea foams white over rusty islands where chunks fallen from the cliffs emerge between the pale pebbles and are worn smooth by the waves. Up in the cliff face, the red-brown mudstone of the Permian Period gives way to the coarser pebble beds and dune sands of the Triassic. In April 2019 we took the children there on an unseasonably hot spring day to discover the end-Permian boundary outcrops in the nudist part of the beach.

The rocks tell us of past climates and tectonic drift too. At Budleigh, the rustiness of the red beds dyed by oxidised iron speak of a far-gone time when this patch of Earth was arid and tropical. Some of the pebbles hidden in the cliffs still bear the coating of varnish bequeathed to them in ancient deserts within the heart of the supercontinent of Pangea. These red rock layers underlie large areas of Devon. The characteristic rusty tan of the county's ploughed fields and the red-tinted torrents that spread ruddy silt across the roads when it rains are the whispers of this ancient world. I am a frequent visitor to Devon and can vouch for the fact that it is far from arid today. After a damp family walk, the hallway can look like a crime scene with brash stains of almost blood-red mud from our boots streaking the pale tiles. It is an image that sometimes draws my mind back to the great dying at the end of the Permian Period that some of these red beds might hold captive, and other times just exasperates me that we never remember to use the door mat.

Closer still to Oxford is the end-Triassic world. On the southern coast of the Bristol Channel, where the gently rolling Quantock Hills meet the muddy sea, the quiet setting of St Audrie's Bay hides a world-class succession of rocks recording this tumultuous piece of Earth's history. Although it is not far from where I grew up, I had never been to St Audrie's Bay before having a scientific interest in it. It was a damp and windy March day last time I visited. The caravans of the holiday parks that crest the cliffs seemed largely deserted and we had the beach almost to ourselves. Where the path down from the car park hits the beach, a turn to the right takes you backwards through time walking deeper into the ancient Triassic Period step by step, stratum by stratum. To your left the rocks grade younger towards the Jurassic. When we arrived, the Jurassic Period was still hiding beneath the tide, so we tramped the pebbled shore to the right round a small promontory, cleaved by an impressive waterfall plunging down the 20-metre cliff-face, and

On the beach at St Audrie's Bay (Somerset, UK), looking west up through the rocks of the Upper Triassic (the Branscombe Mudstone Formation of the Mercia Mudstone Group) in 2022

here found shelter from the whipping westerly breeze. In this part of the beach the green-grey and red striped mudstones of the late Triassic Period have weathered into great westwards-dipping steps in the foreshore. Ladders built to help beachgoers navigate these ancient surfaces have rusted into dilapidation and chromatic sympathy with the reddened rock layers. While we were there, a section of the cliff crashed in with a great rumble, exposing new surfaces, fresh from 200 million years of isolation from the elements.

As the tide sucked back, we ploughed back west into the face of the wind and up through the stratigraphy. To the west of the bay we lost the red beds, and the greys and whites of the shales and limestone started to tell of rising sea levels and a marine environment. At St Audrie's Bay, the harder layers of rock stand proud, weathered out between the mud, and so in many places you climb in steps – taking great strides through geological time. It is estimated that over 70 per cent of species were lost at the end of the Triassic Period. Many of the fossils in these layers do not immediately present themselves to the naked eye, and the remains of the inhabitants of ancient ecosystems remain shy and hidden. But, just around the bay's westerly point, stepping up onto one shaley layer, suddenly the biological evidence of a new Jurassic age is unmissable. Between the greens and browns of the seaweed, the sleek grey of the smooth, slanting rock face is peppered with the fist-sized coiled pressings of ammonite shells. Around the corner, on the beach near the village of Watchet, you can find them the size of car tyres. Similar ammonite pavements are found near the south coast UK town of Lyme Regis, and selling these 'snake-stones' was one way that famous fossil hunter Mary Anning supported herself in the early 1800s. They are evidence of biology rebounding after the end-Triassic crisis. Places like Budleigh Salterton and Lyme Regis are rightly valued as places to escape to on holiday away from the clamour of city life, but hidden within these welcome settings for seaside retreats, unexpected geological drama can be found in the foreshore if you know where to look.

Dramatic as they are, tracking the big five mass extinctions through the strata is not the only way that geoscientists can follow Earth's past patterns of environmental change. Hidden in the fossil record are many smaller overturns and changes in biology that can also be discerned. The big five mass extinction events are part of a continuum with many other smaller extinction events captured as surfaces of lower diversity of life in the strata. But layered over these changes in biology, often caught in the same sediments, are changes in the very chemistry of the rocks that tell of past times and environments. Today geochemists employ a plethora of detailed measurements on the sedimentary rocks chipped out of cliff faces, beaches or other places where they are exposed or drilled and extracted as hundreds of metres of core from the continents or seabeds.

One cold December day in the middle of the global Covid-19 pandemic in 2020, I visited an apparently unremarkable corner of the Shropshire countryside near the village of Prees. Here among rolling farmland, one of my colleagues was leading an international operation to drill deep into the layers of ancient seabed hiding under this modest rural landscape. Written in the biology and chemistry of the strata contained in the over 650 metres of rock core hauled from this well are messages about Earth history spanning from 175–201 million years ago. The pandemic meant that the usual protective clothing required to work on the drill site – hard-toed boots, fire-retardant overalls, helmet and high-vis vest – was augmented by a mask and frequent use of hand sanitiser. Every six hours a new six-metre section of Earth's past was fished out of the hole on a wire and our day was filled with cutting, washing, measuring and describing these new cylinders of rock, sheltering from the rain in an adapted shipping container with our hands and clothes coated with grey Jurassic mud. White shelly layers punctuated the dark grey mudstone background of many of the sections that I logged. In places, slices through oyster-like fossilised Gryphaea danced as a white script on the background of black rock, like hieroglyphs on a chalk board, messages seeing first light for almost 200 million years.

This precious material drilled from Shropshire beside the gentle buzz of traffic along the A49 will be pored over by scientists for many years to come. The fossils present will be studied and logged. Chips of material will be taken at regular intervals down the core and their chemistry determined using many different techniques, each aiming to build up a picture of the state of the Jurassic planet. Analysing the balance between lighter and heavier atoms (isotopes) of carbon captured through the layers will inform us about shifts in the Earth's carbon cycle. Analysis of carbon isotopes in the Jurassic layers of Prees's sister core, drilled at Mochras in west Wales in the late 1960s, and other rocks of similar ages around the world show big swings. We expect to see similar patterns in the new material from Shropshire and these signals tell of changes in the balance between the input fluxes of carbon to the Earth's atmosphere and oceans, like planetary outgassing or wildfires, and sinks pulling carbon out of these reservoirs, like weathering of the continents or the burial of carbon-rich sediments.

These swings in the fingerprint of carbon recorded by sediments are not confined to the Jurassic Period but punctuate the entire geological record. Sometimes the rocks that record them are visually unremarkable and it is only in the lab that the story of carbon cycle turmoil is revealed. At St Audrie's Bay I stood with my feet spanning the fine layers of barnacled and seaweed-festooned mudstones, sandstones and limestones that record a strong swing in Earth's global carbon isotope balance. Without my colleague's expert eye, based on his past sampling and chemical analysis of these rocks, I would have been none the wiser. In some instances, these changes in carbon isotopes are more visible. For example, some shifts in the carbon cycle coincide with widespread (sometimes global) dark bands of organic-rich shale rock laid down on past seabeds and preserved in present-day cliffs or cores capturing a specific 'moment' of geological time. These dark layers are evidence of widespread marine anoxia, when the world's oceans bled out their vital dissolved oxygen and became toxic dead zones. Such anoxia is usually taken as symptomatic of global warming. Building up different and ever more detailed chemical and biological data tracing the changes recorded in the sediments laid down during Earth's history gives us an increasingly complete picture of past environments and their evolution.

Linking these encoded messages to the complexity of causes in the Earth system is then the key challenge. Recent debates about the impacts of climate change on different regions and meteorological phenomena highlight how difficult drawing clear lines from cause to effect can be, even for the present-day planet. Delving back into geological time, things only become more uncertain with more unknowns. However, because of the short tenure of our direct observational records and the complexity of our planet's climate system, understanding how and why Earth's environment changed in the past is arguably a vital component of testing our predictions of how it might evolve in the future, especially against the backdrop of the current 'large scale geophysical experiment' of our prolific carbon emissions. Volcanism is just one of the potential causes of past environmental turmoil. But given the ability of large igneous provinces to mobilise carbon from the Earth's interior periodically through geological history at an almost comparable rate to the current efforts of our species, studying the consequences of these episodes of volcanism perhaps offers us a particularly valuable lens through which to look into Earth's future as well as its past.

Efforts to link up the timing of large igneous provinces with evidence about their contemporaneous environment largely rely on our ability to accurately date their lavas and to match these up with the timescale of changes in fossilised life and ocean and atmospheric chemistry laid out in the sedimentary record. In some rare cases, such as the end-Triassic successions in Morocco and north-western North America, large igneous-province lava flows from the Central Atlantic Magmatic Province interleave with good sedimentary records tying them into the same stratigraphy and thence time succession. But in most cases, matching the records requires relating the age of a lump of lava from one part of the world to a sedimentary sequence in a completely different location. There are still significant error bars and uncertainties associated with this matching. Nonetheless, considerable progress has been made, informed by improvements in dating techniques since the 1980s, and an ever more detailed picture of the relationship between the vast episodes of large igneous province volcanism and the global biosphere and surface environment has begun to develop.

The pattern that emerges is both striking and complex. Four out of the 'Big Five' mass extinctions overlap, at least approximately, in time with large igneous provinces: the late Devonian mass extinction with the Viluy Traps, whose remnants outcrop in modern-day Siberia; the great dying of the end-Permian Period and the vast Siberian Traps; the extinction at the end of the Triassic Period and the fractured Central Atlantic Magmatic Province; and the end-Cretaceous dinosaur annihilation with the Deccan Traps (although there is also compelling and well-publicised evidence of a large meteorite impact at this time too). Only the oldest mass extinction in the record – the Ordovician-Silurian event – has no temporal companion flood-basalt, at least as far as we have yet discovered. But this far back in time, the rock record is patchy and incomplete. Most of the Earth's ocean floor is less than 200 million years old, with older oceanic crust having largely sunk back into the planet's mantle at the world's subduction zones during tectonic processing. Rocks on the continents, too, have been worn by weathering and mashed up by plate movements and tectonic forces and might be so diminished, buried or hidden as to be effectively lost to our science. This leaves open the hypothesis that the Ordovician-Silurian mass extinction could also be twinned in time with a large igneous province – if only we had some other way of detecting the ghost of its presence. But, of course, equally importantly, it might not be. We would like evidence that we can trust of either absence or presence of a flood basalt this far back in time, but in terms of the igneous rock record, the destructive forces of weathering and tectonics leave us with a lack of reliable evidence either way.

As well as the absence of potential culprit volcanism for the Ordovician-Silurian mass extinction, if we only compare the occurrence of flood basalts with the 'Big Five' mass extinctions we miss other things too. Given that there is evidence for more than 24 large igneous provinces through Earth history and only five *major* extinctions, although the threshold to qualify as major is somewhat arbitrarily defined (as loss of 70 per cent of species or more over a span of a few million years), this by simple arithmetic leaves most large igneous provinces with no mass-species slaughter to their name. However, if we expand our examination of the sedimentary record to include more minor periods of biological overturn, carbon isotope excursions and evidence of deoxygenation of the oceans, we find that many of these times of environmental upheaval match up with flood volcanism, at least within the uncertainties in the measurements. The lavas that built the landscapes of northern Skye, Mull, Giant's Causeway and the rest of the North Atlantic Igneous Province erupted about 55–60 million years ago and coincided with a notable perturbation in Earth's carbon cycle recorded as an excursion in the values of the carbon isotopes held in the sediments from their more normal background values. Studies of the archives have estimated average global temperatures to have been 5–8°C warmer. Others, like the Karoo-Ferrar large igneous province erupted in the Jurassic Period – now split between southern Africa and Antarctica – and the mighty

Ontong Java ocean plateau pushed out during the Cretaceous Period, match up with dark shale layers in the rock record that indicate widespread ocean deoxygenation. Still, there are some flood basalts, including the most recent Columbia River Basalt Group and the Early Cretaceous Paraná-Etendeka large igneous province, that seem to lack a definitive association with anything as strikingly out of the ordinary in the sediments. Many reasons for these more subdued responses have been suggested, including their smaller size, lower average rates of magmatism, the configuration of the continents, changes in biology's ability to shunt carbon out of the atmosphere or simply that the climate-response itself was more subtle. These differences highlight important complexities in the association between large igneous provinces and fluctuations in Earth's environment. Nonetheless, in the main the evidence suggests that the Earth system and its dependent biology does not seem to thrive during the time periods associated with these huge outpourings spewing forth from the planet's innards.

Correlation, however, does not necessarily imply causation, especially when uncertainties in the dating of flood basalts and relating their exact timing to the geological record are in play. Although the warming of the planet by emission of carbon dioxide, acidification of the environment by hydrochloric and other volcanic acids or depletion of stratospheric ozone are among the grim menu of possible planetary degradation mechanisms from a large igneous province, there are other potential triggers too. An asteroid impact is now widely accepted to have been key to the demise of the dinosaurs. Other coincidences between evidence of impacts and global change are less readily apparent, but again, compiling a reliable record of previous collisions back into the depths of time is fraught with challenge. The Earth's shifting continents have also been proposed as a trigger for past change as their rearrangements can alter ocean currents, wind patterns and habitat configurations. Feedbacks within the Earth system like those between ice-cap volume and sea level, and global warming and the hydrological cycle, might then amplify these changes into global catastrophe. We would like to be able to understand these triggers and these processes in greater detail, but the challenges of peering back tens to hundreds of millions of years into the past are significant and require us to draw on all possible information in our attempts to decipher the messages in Earth's ancient rocks.

Timing is crucial when it comes to understanding the ways in which a large igneous province might trigger global change. Not only do we need to match up the volcanism with the environmental event, but we would also like to understand how the complex waxing and waning of province-wide eruptions over the million-year lifetime of a flood basalt matches up with the complex cascade of changing carbon isotopes and disappearing biology laid down in the sediments. Unexpectedly, a chance conversation during the first year of my PhD and subsequently a lot of carrying of delicate glassware and heavy batteries up today's active volcanoes turned out to have some bearing on these problems of ancient cause and effect.

Back in 2002, I was lucky enough to be assigned a place on a five-week Europe-wide interdisciplinary atmospheric sciences research workshop in France. One of my colleagues on this course was researching the behaviour of the element mercury in polar environments. Mercury is a very unusual substance. It is the only metal that is liquid at room temperature and ambient pressure. We used to use it in thermometers for school science experiments and I remember being fascinated watching its silver drops dart across the bench and floor when there was a breakage. Tragic industrial incidents like the dumping of mercury compounds into Minamata Bay, Japan between the 1930s and 1960s have demonstrated its terrible toxicity when it gets into food chains, and its use in devices like thermometers is now regulated. Sitting across the table one lunchtime between lectures, talking about polar mercury, the question came up about how much mercury volcanoes emitted and what its chemistry was. I could not immediately answer but promised to take a look for relevant scientific papers when I had the chance.

Checking what had been published about this, it became apparent that, while it looked as if volcanoes were a significant natural source of this potent metal to the atmosphere, there were plenty of gaps in our knowledge and understanding. It was also clear that making good measurements of mercury in volcanic plumes was tricky. Different methods to measure the full chemistry of the emitted mercury involve delicate glassware or large power-hungry boxes of electronics with instructions requesting temperature-controlled, dust-free conditions – challenging in even the most benign volcanic situation. Nonetheless, with other UK colleagues and collaborators from Italy and the US and a considerable amount of cursing as we fiddled with delicate equipment in high winds, acid gases and, on one particularly fraught occasion at the top of Etna, hailstorms and whiteout, we did manage to make a series of mercury emission measurements and considerably improve our understanding of this flighty metal. Importantly we showed that, unlike the other types of metals emitted by volcanoes, mercury is released mainly as a gas. This means that it will not settle out like the metals held in volcanic particles and can be transported long distances from the volcano with a lifetime in the atmosphere of months to years.

At the time, our main interest in these results was whether volcanic mercury emissions were an issue of local pollution or if they contributed to the global background atmospheric concentrations. We were not at the time thinking of mass extinctions and large igneous provinces. But then, in 2012, some Canadian-based colleagues published their measurements of a section of sediments downwind of the Siberian Traps covering the period of their eruption and recording the great dying of the end-Permian extinction. There, coinciding with this huge environmental upheaval, captured in the sediments was a sharp spike in the concentrations of mercury – a chemical whisper of the volcanic exhalations from the Siberian Traps, perhaps.

The prospect of a fingerprint of volcanism in the very sediments that record environmental and biological change is a tantalising prospect. It would help us to see whether the volcanic emissions preceded or post-dated the other signals of carbon-cycle perturbations or species overturn. Since this first measurement, my team and colleagues around the world have run great numbers of sedimentary samples covering most of the geological record's significant events. The end-Permian signal remains one of the sharpest, but there are many other intriguing signals, even if mercury spikes are not ubiquitously present in all time layers where we might expect them. There is still much to work out about what these mercury signatures really mean and how they are captured and preserved. But there have been some promising results. Analysing the mercury in the dusty rock powders harvested from the damp foreshore of St Audrie's Bay, we think that we can pick out pulses in the volcanic activity and gas emissions across the great expanse of the Central Atlantic Magmatic Province. One of these mercury enhancements is in the layers capturing the end-Triassic mass extinction. From what we know about the stratigraphy it seems likely that this mercury was gasped out from ancient lavas now left in great stacks in the Atlas Mountains of present-day Morocco. Approximately 200 million years ago, this mercury escaped the magma and volcanic turmoil and drifted with the breeze north over Pangea. Who knows what precise journey it took through the atmosphere, hydrosphere and biosphere, but finally it got locked into the sediments collecting on the bed of an ancient shallow ocean near what is now the UK. Here it sat held in the rocky layers, since thrust up to border the Bristol Channel, until some of my colleagues took an interest and saw fit to chisel out the slivers of the strata on this quiet beach bordering this stretch of muddy sea. Dried, powdered and heated in a glass crucible, this ancient metallic volcanic breath was finally released to trace out its message across the epochs via the blip in the spectrometer trace skirting across a computer screen in my lab in Oxford.

Although their meanings are not always clear, I am sure that there are many more such epistles from ancient volcanic emanations hidden in sedimentary rocks just waiting to be decoded, further pieces of the puzzle of how our gloriously complex Earth system operates. We do not understand the mercury signals well enough as yet but maybe one day this and other new chemical markers might help us to hunt for volcanism further back in time when the provinces themselves might be lost to us due to weathering and tectonics. Our hope is that searching for mercury spikes combined with other signals will give us new insights into the causes of events like the Ordovician-Silurian mass extinction as well as the more recent geological record.

Between November 2019 and January 2020, the UK news featured many stories and pictures of the unusually intense bushfires raging on the other side of the globe in Australia. By the end of January, more than 100,000 square kilometres of land had been scorched: an area comparable to the whole of England. At least 33 people had been killed and thousands of buildings destroyed. Farm livestock and wildlife also took a brutal hit. Smoke plumes from the fires had clobbered air quality in major cities like Canberra, across southern and eastern Australia and as far away as New Zealand. Satellite images traced these tendrils of sooty black carbon smoke as they stretched over 10,000 kilometres across the Pacific towards South America.

Although directly attributing these fires to climate change is complicated, records show that Australia's temperatures have been edging up and are slightly over 1°C warmer on average now than they were in the 1950s. The patterns and oscillations of the ocean-atmosphere system that controls our weather are complex, but hotter, drier conditions make it easier for fires to spread and grow. While the tragic fires were burning in Australia, far away in Oxford I was working with Chinese collaborators and other co-authors to finalise a paper exploring the variations in fossil charcoal abundance from sediments recording tropical peatland during the end-Permian environmental crisis coincident with the lengthy eruptions emplacing the Siberian Traps. An increase in charcoal abundance in our records documented increased wildfire occurrence and matched up with the carbon-cycle disruption captured in the sediments' isotopes. These layers underlie and coincide with the strata recording the changes in biology that characterise the end-Permian mass extinction. Despite all the complexities of teasing out cause and effect over 250 million years ago as well as in the present day, there was something about this juxtaposition in my consciousness of the phenomenon of widespread wildfires now and during a time of such extensive crisis in Earth's deep past that took my breath away.

Nonetheless, there are important differences between the global change that we are observing today and what we see play out in the geological record during flood-basalt intervals. If, as is often but not ubiquitously suggested, carbon from large igneous provinces is a key player in Earth's episodes of rapid environmental change, then comparing volcanic to human emissions seems an important exercise. We have already discussed earlier how human fossil-fuel emissions are still some significant way off the emissions totals estimated during the full lifetime of a large igneous province episode. Only by burning all known reserves of oil, gas and coal (projected to be about three trillion tonnes of carbon) would we get close to our best guess for the total from a flood basalt event. But the rate at which we are pumping carbon out now is far greater than our best approximations of the average emissions rates made by delving into the geological record of these volcanic monsters. Only during their brief most intense phases might the carbon dioxide degassing rate from flood basalts compare to current human emissions according to the state-of-the-art of our present thinking. Volcanism doesn't just emit carbon, of course, and we are always looking to understand more about the co-emission of haze-seeding sulphur and other powerful acids and reactive elements into the environment, and how this might alter the cascade of environmental effects. That said, we should remember that it is not just the carbon cycle that human industry has pulled out of kilter either. For example, we rely on fertilisers to feed the world, and these fertilisers warp the world's phosphorus and nitrogen cycles into new shapes around our needs. These changes contribute to environmental problems such as low-oxygen coastal dead zones in marine ecosystems which might draw us back again to the geological record and the dark layers recording marine anoxia.

We have certainly seen changes to our planet since the industrial revolution at a pace that is, as far as we can tell from the geological record, unusual compared to what has gone before. The average temperature of the world's atmosphere and oceans is ticking up. Ice caps, sea ice, glaciers and snow cover are all decreasing. Sea levels, ocean acidity and extremes of heat and rainfall are all increasing. Human expansion directly threatens biology too. Humankind takes more and more of the planet's resources for itself, fragmenting and destroying habitats, introducing non-native animals and plants that disrupt ecosystems, spreading pathogens, hunting or culling species directly, and changing global climate. We are nowhere near a mass extinction just yet, with at most a few per cent of catalogued species having been lost in the last few hundred years (remember geologically major extinction are often defined as loss of over 70 per cent of species, albeit over periods of millions of years). Nonetheless, studies of historical data suggest that recent and current extinction rates risk accelerating towards a level that could accomplish the sixth mass extinction in just a few centuries.

It is challenging to compare the pace of extinction over the very different timescales of the geological record and historical data. As discussed above, it is also difficult to solidly attribute cause and effect in terms of the links between large igneous provinces and Earth's past global crises. Nonetheless, we know enough already that we should hear alarm bells echoing down to us through geological time. Flood basalts are a mechanism with the power to release vast, if poorly quantified, amounts of carbon into the atmosphere. In many of these past global experiments we see the carbon cycle reeling in the kicks in the carbon isotope record, and in some cases a catastrophe follows that is of sufficient scale to touch most of life on the planet at the time. Understanding better the messages held in the sedimentary and volcanic rocks recording our planet's past will undoubtedly help us to understand how it works and to make predictions about its future. But I think that contemplating the links between the periods of abrupt change that our planet has undergone in its past and the great stacks of basalt left in Skye, the Columbia River, Siberia and many more places besides, yields a more spiritual lesson of perspective too.

Many of my early adventures in Volcanoland involved travelling the world to far-flung places in search of fuming vents to run my experiments and the violent deposits of past eruptions to study. It was in many ways an education in the power of the Earth, and the stories of the communities and scientists who had lived with and worked to make sense of these volatile forces in the past offered guidance about how to seek scientific inspiration and personal solace in and from their potency. Moving my more recent research focus to large igneous provinces, perhaps the epitome of volcanic might, could be seen as a culmination of this journey. But in it there has also been an odd reversal for me. Throughout history and around the world our instinct as humans has often been to deify and anthropomorphise volcanoes. We have come across some of these tales during our adventures here. The ancient Greek seafarers explained Stromboli's outbursts as arising from the keeper of the winds, Aeolus, being held within the mountain. In Guatemala, the volcano has been personified as the supernatural Juan Noj who owns Santa María and sends ash and flows as vengeance if local populations do not adhere to his wishes. On Hawaii, the volcano goddess Pele remains an important part of the islands' culture. Many similar legends abound around the globe to explain these seemingly living, breathing mountains and their sometimes erratic and destructive nature. We are still nowhere near being able to control volcanoes and likely never will be. But, when faced with the glowing vent of Masaya, the clinking flows of Etna or the explosive columns of Santiaguito or any of the other awe-striking volcanic phenomena that we have met in these pages, as well as a feeling of human fragility, I am now confronted, too, with a reflection of our own power as a species.

It seems to me no coincidence that this growing duality in my sense of the message bursting from the Earth in volcanic fumes and rock coincides with a sense of homecoming and re-centring in terms of the sources of the samples I study. Analysing rocks from St Audrie's Bay in Somerset and Prees in Shropshire, in my lab close by in Oxford, for the vestiges of ancient volcanic mercury transported across the globe during times of prolonged planetary-scale turmoil, brings sharply home a sense of the immense strength of volcanism's Earth-shaping reach. Whatever the details, volcanoes have undoubtedly changed this world and many others besides it. Now *we* have that power too. Gazing into a roiling vent we should not just feel small, but also feel big. Even in the face of volcanic power and fury, we cannot hide from our responsibility for our actions, for each other and for this planet.

Epilogue

I have now spent almost half my life studying volcanoes. It has been quite a journey: ascending escarpments, sliding down ice caps and fighting through forests to try to get a closer look at the molten Earth and to understand it better. I have learnt a great deal about how and why volcanoes erupt and how we can best try to predict what they will do in the future. I have thought a lot about their impacts on the lives and livelihoods of the people living on and around them, and I retain a profound respect for the local scientists and officials whose job it is to try to successfully mediate between the Earth and humanity. However, volcanoes teach us not only about our own place on the planet, but about how our planet came to be the way that it is: a story of the inner Earth creeping beneath our feet; continents dancing through the eons; great tectonic plates crashing into one another, raising mountains in slow motion and pulling themselves apart; the thin blue line of our atmosphere; a fine balance of not too hot and not too cold. The story of Earth as Volcanoland is sometimes a fiery one but it is ultimately a tale of immense fecundity and, as we cast our eyes into the heavens and probe other worlds, there is a sense of ever greater wonder at the many pathways our planet's evolution could have taken or still might take, and the benign beauty among which our species has grown.

Amid all the wonder about our past and present, Volcanoland can also highlight grave concerns about our future. My journey started with the twisted death casts of the victims of Vesuvius buried in Pompeii. Unfortunately, although we have got better at predicting and avoiding them, there have been many such tragedies since AD 79 and, however much our science continues to improve, there will lamentably almost certainly be more in our future – it will be a long time before we understand the full depth and breadth of the processes driving the many styles of volcanic eruptions and how to anticipate them. But studying volcanism, geological environmental crises and mass extinction brings another, deeper sense of foreboding. There is much evidence that we are in the midst of another great era of destruction, and many ecologists are already calling this the sixth major mass extinction. There is no large-scale volcanism this time. There are no great fields of lava being pushed out at rates comparable to all the other volcanoes on land added together, as during the eruptions of the large igneous provinces. There will be no vast trap landscapes for future thinkers to explore and link to this potential sixth great turnover in the fossil record, or indeed a splitting supercontinent or – so far at least – a vast impact crater. If this current mass extinction plays out, it will be alongside the human experiment, and when it's over the Earth's volcanoes will still be here, presiding over whatever planet we leave behind.

Over long timescales, these sentries of Earth's history will likely oversee a return to some sort of balance, pulling together with other forces – the atmosphere, rainfall, weathering and marine bugs – to gradually pull humankind's discarded carbon out of the atmosphere and back into the Earth. This will operate far too slowly to abate current trends, however, and we should heed the great catastrophes of our planet's past. There is no way that the huge gasps of deep gas leaking out from large igneous province magmas could be stopped, but today it is up to us as a species to determine what happens next, what world we will leave behind and how we will be judged as ancestors, however challenging making the right decisions might be. I am as much to blame as anyone. It has not passed me by that my journeys to volcanic places come at a carbon cost and I think ever more carefully about whether trips are truly needed. Increasingly, as well, myself and others are turning our research towards ways volcanism might offer solutions, too, by addressing scientific questions associated with geothermal power and supplies of metals for batteries and the other technologies that we will need to adapt our activities to be less carbon-intensive. I have no illusions about it being easy for any of us.

Volcanoes hold a multiplicity of timescales within them. Processes occurring over millions of years to fractions of a second determine their consequences. They hold in them the human timescale. They figure in our written histories and, for some of us, interact with our personal stories. They change with us over spans of time that we can readily make sense of. But their rocks and other markers in Earth's geological archives are also messengers from the deeper history of our world. They remind us of how short our own existence is, and how much bigger and older our planet is. They require us to imagine places that are lost to us in time or inaccessible in their conditions of temperature or pressure to our frail abilities to truly experience them. These are challenges that I come up against time and time again during my adventures in Volcanoland. It makes my head spin with vertigo sometimes, but learning to think outside the brief realms of human history into the echoing expanses of geological time stretches our intellect and gives us fresh insights and perspectives. When we cast our eye over the Bay of Santorini and imagine how it was hundreds of thousands of years ago, or gaze into the roiling gold of Masaya's lava and imagine its long and unseeable journey from the Earth's depths, or contemplate the dark shades of the maria on the Moon and feel their billion-year ages hit us with our own brevity, we start to think way beyond our very limited lifespans and the immediate reach of our species. Perhaps the more that we can train ourselves to do this – to stand in a landscape, volcanic or otherwise, and think on its deep past, present and future – the better we will be at contemplating the linkages of long-term cause and effect that we must get our heads around to tackle the complex problems our world faces.

In the more immediate future, I hope that I get at least another 20 years exploring Volcanoland. There are many things that we know still need to be solved, and looking back over the history of science, including the past volcanological thinking explored in these pages, reminds us again and again not to imagine that all we think we know already is entirely correct. There will without doubt be more eruptions to study and I hope very deeply that these are stacked firmly towards the scientifically interesting, without the horrors of past human tragedies. There are many efforts globally to try to ensure this is the case. At some stage in the future we will almost certainly see another magnitude eight 'super eruption' like the ancient events at Toba or Yellowstone, and one day another large igneous province will puncture the Earth's crust and flood some corner of the world with basalt. There is no indication at the moment that either will occur during my short sojourn on this planet, for which I am deeply grateful – even though witnessing such an event would help me to answer many fundamental scientific questions.

One thing of which I am sure is that the contradictions of studying volcanoes will continue to hold me rapt. To try to understand them feels both primal and deeply alien. Volcanoes span great swathes of time, cover vast areas and seem to occur throughout Space, and yet they can be sudden, frightening, immediate and hold you entirely in a moment. They tell us many things about Earth's past, its fundamental nature and how it has changed. But as well as drawing this line connecting us to our planet's history, volcanoes are also beacons into Earth's future. Unlike humanity or indeed most other current complex species, we know that volcanoes of some sort will stick around to see out more eons of future geological time, as our planet's 'internal fires' continue to slowly cool. The players will change, of course. Volcanic arcs will come and go. New rifts will open and new mantle hotspots will rise to the fore. The continents will dance to this tune and reconfigure their arrangements. The world's volcanoes will continue to shift and shake and die and renew. It may seem strange but, despite the hazards they present and the fact that I will be long gone, I feel a sadness at a world without the majesty of Etna, the beauty of Villarrica or the understated intrigue of Masaya on it.

Such feelings aside, volcanoes of many different ilks will still be there for future earthlings, whoever or whatever they may be, to explore. Volcanic paths can be rough and rocky and are not universally appealing. Nonetheless I hope that at least some of our successors still have inclination to tread the trails through these restless terrains and to look and wonder at the wide vistas of our planet and the worlds of our solar system and beyond, that a profound understanding of Volcanoland can offer.

The author and Rob Martin, summit of Etna, Italy, 2005

Sources, Notes and Further Reading

The map of volcano-related locations at the front of the book is based on the Gall Stereographic projection with the outlines of the tectonic plates taken from the USGS Earthquake Science Center (usgs.gov/media/images/tectonic-plates-earth).

The geologic timescale at the front of the book is adapted from the US NPS Geologic Resources Inventory (2018) and International Chronostratigraphic Chart v2023/06 from the International Commission on Stratigraphy.

The list of references below is not intended to be a comprehensive review of the scientific literature but rather a way into further reading on some of the subjects covered and reference to the sources for quotations used and specific information discussed etc.

Introduction

P. Francis (1993) *Volcanoes, a planetary perspective* (Oxford University Press).
C.H. Langmuir and W. Broecker (2012) *How to Build a Habitable Planet* (Princeton University Press).
D.M. Pyle (2017a) *Volcanoes – Encounters through the Ages* (The Bodleian Library).
D.M. Pyle (2017b) 'Visions of Volcanoes', *19: Interdisciplinary Studies in the Long Nineteenth Century*, 25.
H. Sigurdsson (1999) *Melting the Earth* (Oxford University Press).

Image on p. 3: Photo credit William Mather. Image on p.9: Taken from the 2nd edition of G.J.P. Scrope's *Considerations on Volcanos* (1862), first published in 1825 by W. Phillips (London), Look and Learn/ Bridgeman Images.

Chapter 1

R. Cole (2018) 'Radiant Radium: A Christmas Gift', *Science Museum Blogs*, 26 December (blog.sciencemuseum.org.uk/radiant-radium-a-christmas-gift).

H. Davy (1828) 'On the Phenomena of Volcanoes', *Transactions of the Royal Society of London*, 118, 241–251.

V. Deparis (2014) 'A history of the global understanding of the Earth', *Comptes Rendus Geoscience*, 346, 275-278. (open archive including a picture of Descartes' model of Earth, https://doi.org/10.1016/j.crte.2014.06.001).

R. Descartes (1644) *Principles of Philosophy* (trans.: V.R. Miller and R.P Miller, 1983, D. Reidel Publishing Company).

A.J.S. McGonigle, C. Oppenheimer, B. Galle, T.A. Mather and D.M. Pyle (2002) 'Walking Traverse and Scanning DOAS Measurements of Volcanic Gas Emission Rates', *Geophysical Research Letters*, 29, 1985.

D.M. Pyle (2017b) op. cit.

H. Rymer, B. van Wyk de Vries, J. Stix, G and Williams-Jones (1998) 'Pit Crater Structure and Processes Governing Persistent Activity at Masaya Volcano, Nicaragua', *Bulletin of Volcanology*, 59, 345–355.

H. Sigurdsson (1999) op. cit.

D. Shastri (2016) 'This bee lives on the edge—of an active volcano' *Science*, 29 July, doi: 10.1126/science.aag0711 (sciencemag.org/news/2016/07/bee-lives-edge-active-volcano).

J. Verne (1864) *Journey to the Centre of the Earth* (Wordsworth Classics 1996 edition).

J.G. Viramonte and J. Incer-Barquero (2008) 'Masaya, the "Mouth of Hell", Nicaragua: Volcanological Interpretation of the Myths, Legends and Anecdotes', *Journal of Volcanology and Geothermal Research*, 176, 419–426.

D.A. Young (2003) *Mind Over Magma: The Story of Igneous Petrology* (Princeton University Press).

Image on p. 20: Photo credit Clive Oppenheimer. Image on p.31: Bridgeman Images. Image on p. 43: Alamy. Map on p. 48: The Pacific tectonic plate has boundaries based on the USGS Earthquake Science Center (usgs.gov/media/images/tectonic-plates-earth) using the Gall Stereographic projection with the volcano locations taken from the Smithsonian Institute, national Museum of Natural History, Global Volcanism Program, Volcanoes of the World database, a catalog of Holocene and Pleistocene volcanoes, and eruptions from the past 12,000 years (volcano.si.edu). The cross-section inset is based on a graphic from the USGS Volcano Hazards Program (usgs.gov/media/images/subduction-juan-de-fuca-plate-beneath-north-american-pla-0). Image on p. 53: Photo credit Katia Pasos.

Chapter 2

T. Anderson (1908) 'The Volcanoes of Guatemala', *The Geographical Journal*, 31, 473–485.

J. Ball (2012) 'October 25, 1902: Santa Maria Erupts!', *AGU Blogosphere Magma Cum Laude*, 26 October (blogs.agu.org/magmacumlaude/2012/10/26/october-24-1902-santa-maria-erupts).

H.C. Berry, K.V. Cashman and C.A. Williams (2021) 'The 1902 Plinian eruption of Santa María volcano, Guatemala: A new assessment of magnitude and impact using historical sources', *Journal of Volcanology and Geothermal Research*, 414, 107167.

J. Dvorak (2015) *The Last Volcano* (Pegasus Books).

G. Eisen (1903) 'The Earthquake and Volcanic Eruption in Guatemala in 1902', *Bulletin of the American Geographical Society*, 35, 325–352.

P. Francis (1993) op. cit.

A.J.L. Harris, W.I. Rose, L.P. Flynn (2003) 'Temporal trends in lava dome extrusion at Santiaguito 1922–2000', *Bulletin of Volcanology* 65, 77–89.

A. Kerr (2020) 'Classic Rock Tours 4. Long Walks, Lost Documents and the Birthplace of Igneous Petrology: Exploring Glen Tilt, Perthshire, Scotland', *Geoscience Canada*, 47, 83–102.

R. Menchú (1984) *I, Rigoberta Menchú* (ed. E. Burgos-Debray, trans. A. Wright, Verso).

A. Rojas (2014) 'Tiempo de recorder todas aquellas historias y leyendas caracteristicos del día de los Santos', *El Palmar TV Blog* (elpalmartvextra.blogspot.com/2014/10/tiempo-de-recordar-todas-aquellas).

J.A.J. Scott, D.M. Pyle, T.A. Mather and W.I. Rose (2013) 'Geochemistry and Evolution of the Santiaguito Volcanic Dome Complex, Guatemala', *Journal of Volcanology and Geothermal Research*, 252, 92–107.

H.C. Sorby (1858) 'On the Microscopical Structure of Crystals, Indicating the Origin of Minerals and Rocks', *Quarterly Journal of the Geological Society*, 14, 453–500.

D. Tedlock (trans.: 1996) *Popol Vuh: The Definitive Edition of the Mayan Book of the Dawn of Life and the Glories of Gods and the Kings* (Touchstone).

R.E. Wolf, R.O. Gomez and W.I. Rose, (2010) 'Geologic Map of Santiaguito Dome Complex, Guatemala', *The Geological Society of America Digital Map and Chart Series 8*.

D.A. Young (2003) op. cit.

R. Waterfield (trans.: 2008) *Plato, Timaeus and Critias* (Oxford University Press). Quotation read in R. Macfarlane (2019) *Underland* (Hamish Hamilton).

Image on p. 56: Photo credit Jeannie Scott. Image on p. 58: Potto A Larcroix/Wikipedia. Image on p. 65: Temple Sepia Times/Getty images.

Chapter 3

A. Boettcher (1975) *Santorini: Portrait of an Island* (Mayer'sche Buchhandlung).

T.H. Druitt, L. Edwards, R.M. Mellors, D.M. Pyle, R.S.J. Sparks, M. Lanphere and B. Barreirio (1999) 'Santorini Volcano', *Geological Society, London, Memoir 19*.

T.H. Druitt, D.M. Pyle and T.A. Mather (2019) 'Santorini Volcano and its Plumbing System', *Elements*, 15, 177–184.

T.H. Druitt, F.W. McCoy and G.E. Vougioukalakis (2019) 'The Late Bronze Age Eruption of Santorini Volcano and Its Impact on the Ancient Mediterranean World', *Elements*, 15, 185–190.

F.A. Fouqué (1879) (trans.: A.R. McBirney, 1998) *Santorini and its Eruptions* (The Johns Hopkins University Press).

P. Francis (1993) op. cit.

P. Nomikou, M.M. Parks, D. Papanikolaou, D.M. Pyle, T.A. Mather, S. Carey, A.B. Watts, M. Paulatto, M.L. Kalnins, I. Livanos, K. Bejelou, E. Simou and I. Perros (2014) 'The Emergence and Growth of a Submarine Volcano: The Kameni Islands, Santorini (Greece)', *Geological Research Journal*, 1–2, 8–18.

P. Nomikou, T.H. Druitt, C. Hübscher, T.A. Mather, M. Paulatto, L.M. Kalnins, K. Kelfoun, D. Papanikolaou, K. Bejelou, D. Lampridou, D.M. Pyle, S. Carey, A.B. Watts, B. Weiß and M.M. Parks (2016) 'Post-eruptive Flooding of Santorini Caldera and Implications for Tsunami Generation', *Nature Communications*, 7, 13332.

M.M. Parks, J. Biggs, P. England, T.A. Mather, P. Nomikou, K. Palamartchouk, X. Papanikolaou, D. Paradissis, B. Parsons, D.M. Pyle, C. Raptakis and V. Zacharis (2012) 'Evolution of Santorini Volcano Dominated by Episodic and Rapid Fluxes of Melt From Depth', *Nature Geoscience*, 5, 749–754.

D.M. Pyle (2017a) op. cit.

R.S.J. Sparks, M.I. Bursik, S.N. Carey, J.S. Gilbert, L. Glaze, H. Sigurdsson and A.W. Woods (1997) *Volcanic Plumes* (Wiley).

D.B. Vitaliano (1973) *Legends of the Earth: Their Geologic Origins* (Indiana University Press).

G. Vougioukalakis, R. Sparks, T. Druitt, D. Pyle, C. Papazachos and M. Fytikas, (2016) 'Volcanic Hazard Assessment at Santorini Volcano: A Review and a Synthesis in the Light of the 2011–2012 Santorini Unrest', *Bulletin of the Geological Society of Greece*, 50, 274–283.

Images on pp. 102, 109 & 118: Author's own photographs. Image on p.116: Photo credit David Pyle.

Chapter 4

J. Brewer (2018) 'Teodoro Monticelli, Vesuvius and Naples' *Blogpost* (brewersblog.org/2018/02/18/teodoro-monticelli-vesuvius-and-naples).

C.G.B. Daubeny (1835), 'Some Account of the Eruption of Vesuvius, Which Occurred in the Month of August 1834, extracted from the manuscript notes of Cavaliere Monticelli, Foreign Member of the Geological Society, and from other sources; together with a Statement of the Products of the Eruption, and of the condition of the Volcano subsequently to it', *Philosophical Transactions of the Royal Society*, 125, 153–159.

A.L. Day and E.S. Shepherd (1913) 'Water and Volcanic Activity', *Bulletin of the Geological Society of America*, 24, 573–606.

M. Di Vito, L. Lirer, G. Mastrolorenzo and G. Rolandi (1987) 'The 1538 Monte Nuovo eruption (Campi Flegrei, Italy)', *Bulletin of Volcanology*, 49, 608–615.

J. Dvorak (2011) 'The Origin of the Hawaiian Volcano Observatory', *Physics Today*, 64, 32-37.

J. Dvorak (2015) op. cit.

M. Edmonds, I.R. Sides, D. Swanson, C. Werner, R.S. Martin, T.A. Mather, R.A. Herd, R.L. Jones, M.I. Mead, G. Sawyer, T.J. Roberts A.J. Sutton and T. Elias (2013) 'Magma storage, transport and degassing during the 2008–10 summit eruption at Kīlauea Volcano, Hawai`i', *Geochimica et Cosmochimica Acta*, 123, 284–301.

T.M. Gerlach (1980) 'Evaluation of Volcanic Gas Analyses from Kīlauea Volcano', *Journal of Volcanology and Geothermal Research*, 7, 295–317.

M.R. Halliday and A.A. Cigna (2006), 'The Grotta Del Cane (Dog Cave), Naples, Italy', *Cave and Karst Science: Transactions of the British Cave Research Association*, 33, 131-136.

M.M. Hirschmann (2006) 'Water, Melting, and the Deep Earth H_2O Cycle', *Annual Review of Earth and Planetary Sciences*, 34, 629–653.

T.A. Mather, M.L.I. Witt, D.M. Pyle, B.M. Quayle, A. Aiuppa, E. Bagnato, R.S. Martin, K.W.W. Sims, M. Edmonds, A.J. Sutton and E. Ilyinskaya (2012) 'Halogens and trace metal emissions from the ongoing 2008 summit eruption of Kīlauea volcano, Hawai`i', *Geochimica et Cosmochimica Acta*, 83, 292–323.

S. Mikhail and E. Füri (2019) On the Origin(s) and Evolution of Earth's Carbon *Elements*, 15, 307–312.

Ovid (trans.: Mary. M. Innes, 1955), *Metamorphoses* (Penguin Books).

D.M. Pyle (2017a) op. cit.

K. Rubin, Hawaii Centre for Volcanology website 'The formation of the Hawaiian Islands' (soest.hawaii.edu/GG/HCV/haw_formation.html)

M.P. Ryan (1988) 'The mechanics and three-dimensional internal structure of active magmatic systems: Kīlauea volcano, Hawaii', *Journal of Geophysical Research: Solid Earth*, 93, 4213–4248.

H. Sigurdsson (1999) op. cit.

C. Scarpati, P. Cole and A. Perrotta (1993) 'The Neapolitan Yellow Tuff — A large volume multiphase eruption from Campi Flegrei, Southern Italy', *Bulletin of Volcanology*, 55, 343–356.

USGS Volcano Hazards Program website for Mauna Loa (usgs.gov/volcanoes/mauna-loa).

D.B Vitaliano (1973) op. cit.

P. Wallace and A.T. Anderson (2000) 'Volatiles in Magmas', *Encyclopedia of Volcanoes*, 149–170 (Academic Press).

S. Winchester (2003) *Krakatoa: The Day the World Exploded* (Viking).

D.A. Young (2003) op. cit.

Image on p. 137: Getty Images. Image on p. 145: Photo credit Clive Oppenheimer. Image on p. 165: Author's own photograph.

Chapter 5

Aeronet Aerosol Robotic Network website (aeronet.gsfc.nasa.gov).

I. Allende (trans.: M. Sayers Peden, 2003), *My Invented Country* (Flamingo).

T. Anderson (1908) op. cit.

S.F. Corfidi (2014) 'The Colors of Sunset and Twilight' NOAA Storm Prediction Center, informal publications (spc.noaa.gov/publications/corfidi/sunset/).

C.R. Darwin (1835) *Darwin's Beagle Diary* (1831–1836) (darwin-online.org.uk).

L. Jaroff (2003) 'More Than a 'Scream': A Blast Felt Round the World', *New York Times*, 9 Dec (nytimes.com/2003/12/09/science/more-than-a-scream-a-blast-felt-round-the-world.html).

T.A. Mather, V.I. Tsanev, D.M. Pyle, A.J.S. McGonigle, C. Oppenheimer and A.G. Allen (2004) 'Characterization and evolution of tropospheric plumes from Lascar and Villarrica volcanoes, Chile', *Journal of Geophysical Research: Atmospheres*, 109, D21303.

T.A. Mather (2018) 'Living with volcanic gases', EGU Blogs (blogs.egu.eu/divisions/gmpv/2018/01/30/living-with-volcanic-gases).

C. Oppenheimer (2003) 'Climatic, environmental and human consequences of the largest known historic eruption: Tambora volcano (Indonesia) 1815', *Progress in Physical Geography: Earth and Environment*, 27, 230 –259.

D.M. Pyle (2017a) op. cit.

A. Robock (2000) 'Volcanic eruptions and climate', *Reviews of Geophysics*, 38, 191-219.

H. Sigurdsson (1999) op. cit.

H. Stommel and E. Stommel (1979) 'The Year Without a Summer', *Scientific American*, 240, 176–187.

G.J. Symons, ed. (1888) *The Eruption of Krakatoa, and Subsequent Phenomena. Report of the Krakatoa Committee of the Royal Society* (Trübner & Co., London).

S. Winchester (2003) op. cit.

A. Witze and J. Kanipe (2014) *Island on Fire: The Extraordinary Story of Laki, the Volcano that Turned Eighteenth-Century Europe Dark* (Profile Books).

Images on pp. 184 & 189: Photo credit David Pyle. Image on p.195: CVO Photo Archives/Wikipedia.

Chapter 6

R.G. Andrews (2021) 'A Massive Subterranean 'Tree' Is Moving Magma to Earth's Surface', Quanta Magazine, 15 Sept (quantamagazine. org/a-massive-subterranean-tree-is-moving-magma-to-earths-surface-20210915).

J. Biggs, I.D. Bastow, D. Keir and E. Lewi (2011) 'Pulses of Deformation Reveal Frequently Recurring Shallow Magmatic Activity Beneath the Main Ethiopian Rift', *Geochemistry, Geophysics, Geosystems*, 12, Q0AB10.

W.W. Bishop, ed. (1978) *Geological Background to Fossil Man: Recent Research in the Gregory Rift Valley, East Africa* (Geological Society, London, Special Publications, 6).

D. Bressan (2018) 'The Discovery Of The Great Rift Valley, Where Africa Is Splitting In Two', Forbes, 3 April.

P. Briggs (2009) *Ethiopia*, Bradt Travel Guides.

S. Brune, S.E. Williams and R.D. Müller (2017) 'Potential Links Between Continental Rifting, CO_2 Degassing and Climate Change Through Time', *Nature Geoscience*, 10, 941–946.

W. Buckland (1840) 'Memoir on the evidences of glaciers in Scotland and the north of England', *Proceedings of the Geological Society*, 3, 332-337, 345-348 cited in D.R. Oldroyd (2002) op. cit.

M.R. Burton, G.M. Sawyer, D. Granieri (2013) 'Deep Carbon Emissions from Volcanoes', *Reviews in Mineralogy and Geochemistry*, 75, 323–354.

T.C. Chamberlin (1899) 'An Attempt to Frame a Working Hypothesis of the Cause of Glacial Periods on an Atmospheric Basis', *The Journal of Geology*, 7, 545-584.

Geological Society of London, 'Plate Tectonic Stories: East African Rift Valley, East Africa' (geolsoc.org.uk/Policy-and-Media/Outreach/Plate-Tectonic-Stories/Vale-of-Eden/East-African-Rift-Valley) .

P. Gouin (1979) *Earthquake History of Ethiopia and the Horn of Africa* (International Development Research Centre).

J.A. Hunt, A. Zafu, T.A. Mather, D.M. Pyle and P.H. Barry (2017) 'Spatially Variable CO₂ Degassing in the Main Ethiopian Rift: Implications for Magma Storage, Volatile Transport and Rift-Related Emissions', *Geochemistry, Geophysics, Geosystems*, 18, 3714–3737.

W. Hutchison, T.A. Mather, D.M. Pyle, J. Biggs and G. Yirgu (2015) Structural controls on fluid pathways in an active rift system: A case study of the Aluto volcanic complex, *Geosphere*, 11, 542–562.

W. Hutchison, R. Fusillo, D.M. Pyle, T.A. Mather, J. Blundy, J. Biggs, G. Yirgu, B.E. Cohen, R. Brooker, D.N. Barfod and A.T. Calvert (2016) 'A Pulse of Mid-Pleistocene Rift Volcanism in Ethiopia at the Dawn of Modern Humans', *Nature Communications*, 7:13192.

P. Kent (1978) 'Historical Background: Early Exploration in the East African Rift – The Gregory Rift Valley', *Geological Society, London, Special Publications*, 6, 1–4.

E. Laird (2013) *The Lure of the Honeybird: The Storytellers of Ethiopia* (Polygon).

H. Lee, J.D. Muirhead, T.P. Fischer, C.J. Ebinger, S.A. Kattenhorn, Z.D. Sharp and G. Kianji (2016) 'Massive and Prolonged Deep Carbon Emissions Associated with Continental Rifting', *Nature Geoscience*, 9, 145–149.

D.R. Oldroyd (2002) 'The Glaciation of the Lake District' in *Earth, Water, Ice and Fire: Two Hundred Years of Geological Research in the English Lake District*, Geological Society Memoir no. 25, 255–269.

Oxford University Museum of Natural History, 'Learning more … William Buckland' (oum.ox.ac.uk/learning/htmls/buckland.htm).

H. Sigurdsson (1999) op. cit.

C.A. Suarez, M. Edmonds, A.P. Jones (2019) 'Catastrophic Perturbations to Earth's Deep Carbon Cycle', *Elements*, 15, 301-306.

C. Summerhayes (2017) 'Blowing hot and cold', *Geoscientist Online* (geolsoc. org.uk/Geoscientist/Archive/May-2017/Blowing-hot-and-cold).

US EPA 'Greenhouse Gas Equivalencies Calculator' (epa.gov/energy/greenhouse-gas-equivalencies-calculator).

J.C.G. Walker, P.B. Hays and J.F. Kasting (1981) 'A Negative Feedback Mechanism For the Long-Term Stabilization Of Earth's Surface Temperature', *Journal of Geophysical Research*, 86, 9776-9782.

S.R. Weart (2008) *The Discovery of Global Warming* (2nd edition, Harvard University Press).

F. Williams and P. Mohr (2011) *Ethiopia's Rift Valley, Its Geology and Scenery* (Millbrook Nova Press).

Images on pp. 220 & 226: Photo credit David Pyle. Images on pp. 230–1, 238: Photo credit William Hutchison. Image on p. 232: Author's own photograph.

Chapter 7

I. Allende (2003) op. cit.

M. Anand, R. Tartèse and J.J. Barnes (2014) 'Understanding the Origin and Evolution of Water in the Moon Through Lunar Sample Studies', *Philosophical Transactions of the Royal Society A*, 372:2013025420130254.

S.E. Braden, J.D. Stopar, M.S. Robinson, S.J. Lawrence, C.H. van der Bogert and H. Hiesinger (2014) 'Evidence for Basaltic Volcanism on the Moon Within the Past 100 Million Years', *Nature Geoscience*, 7, 787–791.

P.K. Byrne and S. Krishnamoorthy (2022) 'Estimates on the Frequency of Volcanic Eruptions on Venus', *Journal of Geophysical Research: Planets*, 127, e2021JE007040.

The Cosmic Cast (2021) 'How did astronauts explore the Moon?' (youtube.com/watch?v=XWQTAPJC_a0).

N.M. Curran, K.H. Joy, J.F. Snape, J.F. Pernet-Fisher, J. D. Gilmour, A.A. Nemchin, M.J. Whitehouse and R. Burgess (2019) 'The early geological history of the Moon inferred from ancient lunar meteorite Miller Range 13317', *Meteoritics and Planetary Science*, 54, 1401-1430.

P. Francis (1993) op. cit.

P. Geissler (2015) 'Cryovolcanism in the Outer Solar System', *The Encyclopedia of Volcanoes*, 763-776 (2nd edition, Academic Press).

J.W. Head, L. Wilson, A.N. Deutsch, M.J. Rutherford and A.E. Saal (2020) 'Volcanically Induced Transient Atmospheres on the Moon: Assessment of Duration, Significance, and Contributions to Polar Volatile Traps', *Geophysical Research Letters*, 47, e2020GL089509.

R.R. Herrick and S. Hensley (2023) 'Surface changes observed on a Venusian volcano during the Magellan mission', *Science*, 379, 1205–1208.

J.D. Hofgartner, S.P.D. Birch, J. Castillo, W.M. Grundy, C.J. Hansen, A.G. Hayes, C.J.A. Howett, T.A. Hurford, E.S. Martin, K.L. Mitchell, T.A. Nordheim, M.J. Poston, L.C. Prockter, L.C. Quick, P. Schenk, R.N. Schindhelm, O.M. Umurhan (2022) 'Hypotheses for Triton's plumes: New analyses and future remote sensing tests', *Icarus* 375, 114835.

J. Huang, S. Seager, J.J. Petkowski, S. Ranjan, and Z. Zhan (2022) Assessment of Ammonia as a Biosignature Gas in Exoplanet Atmospheres, *Astrobiology*, 22, 171–191.

L. Kaltenegger (2017) 'How to Characterize Habitable Worlds and Signs of Life', *Annual Review of Astronomy and Astrophysics*, 55, 433–85.

J.F. Kasting and D. Catling (2003) 'Evolution of a Habitable Planet', *Annual Review of Astronomy and Astrophysics*, 41, 429–63.

T.A. Mather et al. (2004) op. cit.

C. Morgan, L. Wilson and J.W. Head (2021) 'Formation and Dispersal of Pyroclasts on the Moon: Indicators of Lunar Magma Volatile Contents', *Journal of Volcanology and Geothermal Research*, 413, 107217.

NASA Earth Observatory (2015) Historic rain around the Atacama Desert led to devastating floods and mud flows. Image of the Day for April 11 (earthobservatory.nasa.gov/images/85685/flooding-in-the-chilean-desert).

NASA Astromaterials Acquisition and Curation Office, 'Lunar Rocks and Soils from Apollo Missions' (curator.jsc.nasa.gov/lunar).

D.H. Needham and D.A. Kring (2017) 'Lunar Volcanism Produced a Transient Atmosphere Around the Ancient Moon', *Earth and Planetary Science Letters*, 478, 175–178.

S.J. Peale, P. Cassen, and R.T. Reynolds (1979) Melting of Io by Tidal Dissipation, *Science*, 203, 892-894.

G. Schubert, R.E. Lingenfelter and S.J. Peale (1970) 'The morphology, distribution, and origin of lunar sinuous rilles', *Reviews of Geophysics*, 8, 199–224.

I. Todd (2020) 'Interview with Voyager scientist Linda Morabito', *BBC Sky at Night Magazine*, 6 July (skyatnightmagazine.com/space-missions/volcano-jupiter-moon-io-interview-voyager-linda-morabito).

A. von Humboldt (1853, trans.: Vera M. Kutzinski, 2010), 'About an Attempt to Climb to the Top of Chimborazo', *Atlantic Studies*, 7, 91–211.

C.A. Wood (2019) 'Before Apollo: The scientists who discovered the Moon', Sky & Telescope, 9 July (skyandtelescope.org/astronomy-resources/before-apollo-scientists-who-discovered-moon).

A. Wulf (2015) *The Invention of Nature: The Adventures of Alexander von Humboldt, the Lost Hero of Science* (John Murray).

F. Wunderlich, M. Scheucher, M. Godolt, J.L. Grenfell, F. Schreier, P. C. Schneider, D. J. Wilson, A. Sánchez-López, M. López-Puertas, and H. Rauer (2020) 'Distinguishing between Wet and Dry Atmospheres of TRAPPIST-1 e and f', *The Astrophysical Journal*, 901, 126.

Images on pp. 255 & 275: Photo credit David Pyle. Image on p. 258: NASA Image Collection/Alamy. Image on p. 262: Geography and Map Division, Library of Congress. Image on p.270: NASA Image Collection.

Chapter 8

A.D. Barnosky, N. Matzke, S. Tomiya, G.O.U. Wogan, B. Swartz, T.B. Quental, C. Marshall, J.L. McGuire, E.L. Lindsey, K.C. Maguire, B. Mersey and E.A. Ferrer (2011) 'Has the Earth's sixth mass extinction already arrived?', *Nature*, 471, 51–57.

BBC news (2020) 'Australia fires: A visual guide to the bushfire crisis', 31 January (bbc.co.uk/news/world-australia-50951043).

P. Brannen (2017) *Ends of the World: Volcanic Apocalypses, Lethal Oceans and Our Quest to Understand Earth's Past Mass Extinctions* (Oneworld).

G. Ceballos, P.R. Ehrlich, A.D. Barnosky, A. García, R.M. Pringle and

T.M. Palmer (2015) 'Accelerated modern human–induced species losses: Entering the sixth mass extinction', *Science Advances*, 1, e1400253.

Y. Cui, M. Li, E.E. van Soelen, F. Peterse and W.M. Kürschner (2021) 'Massive and rapid predominantly volcanic CO_2 emission during the end-Permian mass extinction', *Proceedings of the National Academy of Sciences*, 118, e2014701118.

E. Dunne (2018) 'Patterns in Palaeontology: How Do We Measure Biodiversity in the Past?' *Palaeontology Online*, 8, 1–9.

R. Garwood (2014) 'Life as a Palaeontologist: Palaeontology for Dummies, Part 2', *Palaeontology Online*, 4, 1–10.

A. Geikie (1897) *The Ancient Volcanoes of Great Britain* (Macmillan).

S. Hesselbo et al. (2023) 'Initial results of coring at Prees, Cheshire Basin, UK (ICDP JET Project); towards an integrated stratigraphy, timescale, and Earth System understanding for the Early Jurassic', *Scientific Drilling*, 32, 1–25.

R. Lindsey (2023) 'Climate Change: Atmospheric Carbon Dioxide' *NOAA Climate.gov News & Features*, 12 May (climate.gov/news-features/ understanding-climate/climate-change-atmospheric-carbon-dioxide).

T.A. Mather and A. Schmidt (2021) 'Environmental Effects of Volcanic Volatile Fluxes From Sub-Aerial Large Igneous Provinces', in R.E. Ernst, A.J. Dickson, A. Bekker eds., *Large Igneous Provinces: A Driver of Global Environmental and Biotic Changes*, AGU Geophysical Monograph 255, 103–116.

G.R. McGhee, M.E. Clapham, P.M. Sheehan, D.J. Bottjer, M.L. Droser (2013) 'A new ecological-severity ranking of major Phanerozoic biodiversity crises' *Palaeogeography, Palaeoclimatology, Palaeoecology*, 370, 260-270.

C. McGlade and P. Ekins (2015) 'The geographical distribution of fossil fuels unused when limiting global warming to 2°C', *Nature*, 517, 187–190.

NASA 'How Do We Know Climate Change Is Real?' (climate.nasa.gov/ evidence/).

Oxford University Museum of Natural History, 'Learning more … The statues in the court' (oum.ox.ac.uk/learning/htmls/ statues.htm).

L.M.E. Percival, B.A. Berguist, T.A. Mather and H. Sanei (2021) 'Sedimentary mercury enrichments as a tracer of Large Igneous Province volcanism', in R.E. Ernst, A.J. Dickson, A. Bekker eds., *Large Igneous Provinces: A Driver of Global Environmental and Biotic Changes*, AGU Geophysical Monograph 255, 247-262.

J. Phillips (1860) 'Life on the Earth: Its Origin and Succession' (Macmillan)

G. Racki (2020), 'Volcanism as a Prime Cause of Mass Extinctions: Retrospectives and Perspectives', in T. Adatte, D.P.G Bond and G Keller eds., *Mass Extinctions, Volcanism, and Impacts: New Developments*, Geological Society of America: Special Paper 544, 1–34.

G. Racki (2021), 'The Discovery of the Siberian Traps Large Igneous Province in 1870s, and the Leading role of Aleksander Czekanowski', *Proceedings of the Geologists' Association*, 132, 369–391.

D.M. Raup and J.J. Sepkoski (1982) 'Mass Extinctions in the Marine Fossil Record', *Science*, 215, 1501-1502.

D.M. Raup (1995) 'The Role of Extinction in Evolution Tempo and Mode', in W.M. Fitch and F.J. Ayala eds., *Evolution Genetics And Paleontology 50 Years After Simpson*, National Academy of Sciences (US), National Academies Press (US), 109-124.

R. Revelle and H.E. Suess (1957) 'Carbon Dioxide Exchange between Atmosphere and Ocean and the Question of an Increase of Atmospheric CO_2 During the Past Decades', *Tellus* 9, 18–27.

H.H. Svensen, D.A. Jerram, A.G. Polozov, S. Planke, C.R. Neal, L.E. Augland and H.C. Emeleus (2019) 'Thinking about LIPs: A Brief History of Ideas in Large Igneous Province Research', *Tectonophysics* 760, 229–251.

S.R. Weart (2008) op. cit.

A. Wulf (2015) op. cit.

Y. Zhang, M. Pagani, Z. Liu, S.M. Bohaty and R. DeConto (2013) 'A 40-Million-Year History of Atmospheric CO_2', *Philosophical Transactions of the Royal Society A*, 371, 20130096.

Image on p. 284: Author's own photograph. Map on p. 287 of the large igneous provinces is based on Figure 1 in S. E. Bryan and L. Ferrari (2013) 'Large igneous provinces and silicic large igneous provinces: Progress in our understanding over the past 25 years', *Geological Society of America Bulletin*, 125, 1053–1078. The map is based on the Gall Stereographic projection and drawn by Stephen Dew. Image on p.298: Fig 4. from J. Phillips (1860) op. cit. Image on p.302: Author's own photograph.

Epilogue

Image on p. 322: Photo credit David Pyle.

Acknowledgements

Writing a book like this is not something I thought that I would ever attempt. Thank you to Matthew Turner of RCW Literary Agency for suggesting that I try, and for being so instrumental in helping to mould the initial stream of consciousness into something (hopefully) more readable. Thank you, too, to Jon Appleton, Richard Beswick, David Bamford and the rest of the team at Little, Brown Book Group and to Eden Railsback, John Glynn and the team at Hanover Square Press for believing in the project and all their efforts to help improve it. David Pyle, Rachel Hinton, Bettina Bildhauer, Erin Saupe, Claire Nichols and Atalay Ayele provided invaluable feedback on parts or all of the manuscript for which I am incredibly grateful. Others like Xochilt Hernandez Leiva, Felicity Mather and Charlotte Deane read parts or all of earlier drafts and offered much needed encouragement to keep going. Many other family members, friends and colleagues also provided support in different ways during the journey, especially when the going got tough – enormous thanks to you all.

These are just some of my adventures through Volcanoland so far. There are many other paths through it, and many yet to tread, and this is by no means an attempt at a complete account. There are very many important and spectacular volcanic regions of our planet that I have yet to visit. This account is clearly framed by where I have been, and I am very aware that this is skewed towards Europe and the Americas to date. I have not yet personally done fieldwork in, for example, the very volcanically active western part of the Pacific ring of fire but would certainly jump at the chance to fill in this and the many other gaps in my knowledge of volcanic places. I am also very conscious that my Northern European perspective reflects my origins and education. Time was too short to explore all perspectives as fully as I would have liked at this stage but I always aim to keep learning. There are also many and significant aspects of volcano science that I was not able to fit into these pages, and again the focus of what I have written about in some places covers the science that we were undertaking during the field campaigns I describe at the expense of other equally or more exciting achievements. I hope that I have managed to chronicle something of the breadth and depth of collective efforts in volcanology past and present, but it is far too rich an endeavour to capture it all.

Other people were, of course, present during all the fieldwork described here. I am incredibly privileged to have worked with a large and talented group of scientists during my adventures and to avoid a jungle of names within these pages I decided to limit specific mentions to some of our in-country collaborators. Nonetheless, hidden within these accounts (often in multiple places) is the presence of: David Pyle, Clive Oppenheimer, Andrew Allen, Andrew McGonigle, Ken Sims, Evgenia Ilyinskaya, Matt Watson, Jeannie Scott, Jon Stone, Rudiger Escobar Wolf, Sandro Aiuppa, Emanuela Bagnato, Seb Watt, Rob Martin, Giovanni Chiodini, Evi Nomikou, Michelle Parks, Mel Witt, Roland von Glasow, Tjarda Roberts, Adam Durant, Vitchko Tsanev, Jeff Sutton, Tamar Elias, Jim Kauahikaua, John Catto, Gezahegn Yirgu, Will Hutchison, Karen Fontijn, Fekadu Aduna, Jason Permenter, Victoria Martin, Ben Mason, Mike Rampey, Guillermo Chong, Claire Witham, Sara Barsotti, Wilfried Strauch, Rachel Whitty,

Hilary Francis, Guðrún Halla Tulinius, Xochilt Hernandez Leiva, Steve Hesselbo, Conall Mac Niocaill, Stuart Robinson, Joost Frieling, Isabel Fendley, Asri Indraswari, Alice Paine, Jean-François Smekens, Mike Burton, Terry Plank, Giuseppe Salerno, Nicole Bobrowski, Sandro La Spina and many more. There are many others still who have collaborated on the science that has allowed me to write this book. Some of you are named as authors or co-authors in the references contained in the notes pages but again there were too many to include and I take this opportunity to thank all my scientific collaborators, especially the research students and early career scientists who have trusted in me for mentorship – and from whom I have often learnt more than vice versa – and colleagues in big projects like COMET (led by Tim Wright/Barry Parsons), Riftvolc (led by Kathy Whaler/Juliet Biggs), STREVA (led by Jenni Barclay), V-PLUS (led by Anja Schmidt), Mantle Volatiles and V-ECHO. Scientific work like mine also requires funding and the fact that this often ultimately comes from the public purse is something that I do not wear lightly even if our findings do not always turn out to be as profound or as useful as we might hope. My work described in these pages was funded by the NERC, the ERC, the Royal Society, the NSF, the Leverhulme Trust, a UNESCO/L'Oréal UK and Ireland Women in Science Award, the ICDP, the Boise Trust Fund (University of Oxford), the Centre of Latin American Studies (Cambridge University), University College (Oxford) and others. Further, there are many scientific institutions and hazards management agencies that have been vital to my work around the world and whose responsibility for managing volcanic hazards I am forever in awe of, for example: INGV (Italy), INETER (Nicaragua), USGS (USA), IMO (Iceland), INSIVUMEH (Guatemala), SERNAGEOMIN (Chile) and many others besides.

Lastly, there are many people without whom I would not be the person I am today to write this book. Heartfelt thanks to my parents for setting me on my way and being an endless source of love and support in good times and bad, and to my sister, who I hope will recognise much we share in these pages. As a child and into adulthood I was lucky enough to be very close to three of my grandparents and to hear much about my grandfather who had already passed. Amongst many things, they brought me back my first volcanic rock from Etna and each taught me so much about the power of recounting stories (sometimes repetitively!) and I think about them very often even now that they are all gone. Many thanks, too, to wider family and friends for your support and for putting up with my frequent bouts of self-doubt, my general geekiness and many a spontaneous and sometimes likely unneeded ramble about something scientific and/or volcanic over the years. You know who you are and I am incredibly fortunate to have you all. Learning never stops but it is not something that you do alone. I am immensely grateful to the many teachers, at school, university and beyond, who have influenced me over the years, there are too many to name but I would not be here without you. Deep gratitude, too, to all the people who have helped me to chart being a parent as well as a volcanologist, especially our wider families and friends, nursery staff, school teachers and the team on Kamran's ward. This book is dedicated with love to David, who has been, and is, my co-adventurer in so many different ways and, above all, to Alice and Dominic and your future adventures *whatever* they might be.

Index

About the Author

Tamsin Mather is a British volcanologist. She is Professor of Earth Sciences at the Department of Earth Sciences, University of Oxford and a Fellow of University College, Oxford. She was born and raised in Bristol, UK and has masters degrees in Chemistry and the History and Philosophy of Science from St John's College, University of Cambridge. She completed her PhD in 2004 on the 'Near-source chemistry of tropospheric volcanic plumes' in the Department of Earth Sciences also at the University of Cambridge. Before joining Oxford she was a NERC fellow at the Parliamentary Office of Science and Technology and a Royal Society Dorothy Hodgkin research fellow. She has won numerous awards for her scientific work including the Royal Society Rosalind Franklin Award and election to the Academia Europaea and as a Geochemistry Fellow. She regularly participates in events promoting the public understanding of science and TV, radio and podcasts, including *The Infinite Monkey Cage* on BBC Radio 4.